EX LIBRIS
John W. Lane

KOOKABURRA

KOOKABURRA

the most compelling story in Australia's aviation history

Research by Dick Smith
Story by Pedr Davis

LANSDOWNE PRESS
Sydney • Auckland • London • New York

Published by Lansdowne Books
176 South Creek Road, Dee Why West, NSW, Australia 2099
First published 1980
© Copyright Paul Hamlyn Pty Limited
Produced in Australia for the Publisher by
HEL Productions
8 Cliff Street, Milsons Point, NSW 2061
Edited by Helen Lesnewski
Designed by Robyn Peacock-Smith
Typeset by Netan Pty Ltd
Printed by Kyodo-Shing Loon Printing Industries Pte Ltd
112 Neythal Road, Jurong Town, Singapore 22

National Library of Australia Cataloguing-in-Publication Data
Davis, Pedr Llewellyn.
    kookaburra, Australian aviation's most compelling story.

ISBN 0 7018 1357 1

1. Anderson, Keith, ca. 1899-1929. 2. Air pilots —
Biography. I. Smith, Dick. II. Title.

629.13 092 4

Foreword

Australians are properly and fiercely proud of our national aviation history.

Australian names and Australian achievements are a feature in the world records of this important human endeavour, which more than any other has united peoples of all continents.

The skies above the mighty oceans and seas which wash our shores were tamed by Australian aviators. Pioneering flights of Kingsford Smith and Ulm, P.G. Taylor, Bert Hinkler and Ross and Keith Smith were the foundation of the great 'air bridges' between Australia and sister continents and of the end of the tyranny of distance.

Modern discovery and development of our great country is synonymous with the aeroplane. The early aviators — big men in tiny craft — made their mark indelibly in opening up new frontiers in a land of pioneering and pioneers.

In the slipstreams of their Avros, BE 2Es and, by our standards today, other primitive machines, a tradition of Australian air-mindedness emerged with important manifestations for us all, including the establishment of the unique Royal Flying Doctor Service and the oldest airline in the English-speaking world — Qantas.

The wealth of historical events described in this book formed the background to the early initiatives to harness the potential of the aeroplane for national development. Enter the airlines!

Between 1919 and 1927 at least sixteen passenger and air mail carriers began operations, but only two — Qantas, formed in 1920, and Western Australian Airways, formed in 1921 — survived the decade. The airways held promise but offered no guarantees.

Joy rides — at three guineas each — and air taxi services were the cornerstone and the fledgling Qantas began to stretch its wings, with its first scheduled services, only after 1922. The flight, from Charleville to Cloncurry, over a distance of 577 miles, was flown in two stages to avoid the searing Queensland summer heat with a payload of 400 pounds of cargo and 160 letters.

Imagine the feelings of the first passenger, Alexander Kennedy, as the pilot, Hudson Fysh, nursed along the Armstrong Whitworth FK8 biplane on that flight at an altitude of 4,000 feet. Fifty-three years earlier the grazier had struggled for months by bullock waggon across the same route he would cover that day in just three and a quarter hours!

Aviation's dreams were starting to be realized and its promises to be fulfilled. On May 9th, 1928, exactly one month before Kingsford Smith's *Southern Cross* arrived in Brisbane at the end of its epic crossing of the Pacific, Qantas introduced the first daily air service in Australia.

Australian aviation today, symbolized by modern, hard-working jets serving the nation with safe, efficient, economical and regular services at home and overseas, is the fascinating perspective for the recollections of our pioneers of the skies just 60 short years ago.

by the Chairman of Qantas, Sir Lenox Hewitt

Preface

The circumstances surrounding the loss of the aeroplane *Kookaburra* and the deaths of the two aviators, Anderson and Hitchcock, in the Tanami Desert in 1929, received enormous publicity which set off a spate of rumours, hysteria and emotion.

The story has been resurrected frequently in published articles, with many inaccuracies over the past 50 years.

During the last 10 years an interest in relocating the remains of the aircraft has led to numerous search expeditions, some inspired by genuine historical interest and others by wild stories, reminiscent of 'Lasseter's Reef', that a member of the 1929 truck party which recovered the bodies, had found gold-bearing rock nearby.

It remained for a young aviation enthusiast and self-made electronics millionaire, Dick Smith, to finally locate the skeletal remains of the *Kookaburra* in 1978, ravaged by the passing 49 years and periodic bush-fires.

For the human dynamo, Dick Smith, the search became a challenge and an obsession. His research and story in this book are the most factual I have read, and in themselves are an epic of determination and endurance against the most discouraging odds.

For me the finding of the *Kookaburra* in the lonely desert with the two aviators dead from thirst, was a grim reminder of the tragic consequences of the reckless haste and the taking of unnecessary risks in flying even in those days.

The final prime cause of the tragedy was not the faulty compass, as stated in the official report of the Committee of Inquiry, but the engine failing again while flying over hundreds of kilometres of uninhabited arid desert. If the engine had not failed when it did, they would have found their way safely to Wyndham.

by Lester Brain, AO, AFC.

Dedication

TO MY WIFE, PIP
without whose enthusiastic support
I would not have been able to achieve
what I have in life.
DICK SMITH

Acknowledgements

FROM DICK SMITH
Many people who were personally involved with events at the time of the loss of the *Kookaburra* have helped me in the preparation of this book. Grateful thanks are extended to them all including Lester Brain, A.O., the late Charles Eaton, Sir Norman Brearley, Mrs Bon Tate and Mrs Pretoria Bliss.

I would also like to thank Miss Sally Douglas for kindly making available her father's diary.

For helping with the historical facts, I would like to thank Ted Wixted of the Queensland Museum and John Haslett in Darwin.

For details of the searches for the gold-reef, I thank Rod Liebernecht, Max Morris and Jim Love — all of Adelaide.

Of course it is impossible to mention all people who helped. However I would be amiss in not especially thanking the following:

Messrs Ralph Hayes, David Poynter, Garry Crapp, Gerry Nolan, Charles Eaton (Jnr), Joe Drage, Arthur Whittaker, Norman Bignell, Len Bell, Keith Locke, Ron Berg and Colonel Hatfield.

I would also like to thank the Mercedes Benz Company for the loan of their magnificent Unimog four wheel drive truck.

FROM THE PUBLISHER
Thanks are extended to the following people for their assistance and contributions:

Sir Lenox Hewitt, Chairman of Qantas Airways Limited; Alan Hunt, Tooths Brewery Museum; Robert Lasseter; Bob Piper, Defence Department in Canberra.

Publisher's note

Metric measures of weight and distance have been used in this publication, however where a direct quote contains an imperial amount, a footnote has been added to give the approximate metric conversion. In addition, if an amount relates to a series of statements given at an inquiry for example, the figure is given in the imperial value with the approximate metric equivalent in brackets.

Further, monetary values are stated in pounds (£), since there is no accurate conversion to present-day dollars. In most cases an approximate value in dollars has been given, based on today's purchasing power in comparison with that of about 50 years ago.

Contents

The born loser

The silver and blue *Southern Cross* glistened in the sun as Charles Kingsford Smith brought the mono-plane low over Sydney's Mascot aerodrome.

With Charles Ulm and two American companions, Smithy stepped out to a tumultuous welcome from a crowd estimated at 300,000. The aviators had just completed the first ever flight across the Pacific Ocean.

Somewhere in that crowd stood a tall, taciturn man. His fiancée, Bon Hilliard, stayed with him, her hand on his arm. The adulation and welcoming speeches held for him only the bitter taste of disappointment. Taking a deep breath, Keith Vincent Anderson strode through the crowd. He solemnly shook Kingsford Smith's hand, nodded to Ulm, then hurried into the obscurity of the thronging mob.

Equally unhappy was a small, square-jawed mechanic who spoke with the drawl of a man who had spent his youth in Western Australia. H.S. Hitchcock, known to his friends as Bob, was even more upset than Anderson. The records do not show if he greeted the jubilant aviators at Mascot, but this seems unlikely. Hitchcock later broke into the official reception and handed Kingsford Smith a writ, claiming the sum of £1,000 (this is estimated to be worth $40,000 in terms of today's purchasing power), as compensation for his work in the venture.

Anderson's disappointment also resulted in court action. Neither lawsuit succeeded. However in an unexpected gesture, Smithy (as he was widely called) and Ulm handed Anderson a gift of £1,000. Smithy also signed a promissory note for a further £300. This act of generosity proved incredibly costly, for it sparked a chain of events still unfinished at the time of this book's production. The developments in this unusual story have accumulated in a file kept in aviation enthusiast Dick Smith's office. That file is now almost one metre thick, with more material being added continually.

The story relates the saga of two men — Anderson and Hitchcock. They forfeited their lives while trying to find Kingsford Smith and Ulm when they became lost in the rugged north-west of Australia. Their misfortune grew to become one of the most compelling chapters in Australian aviation history.

•• •• ••

During the early 1920s, Kingsford Smith, Keith Anderson and Bob Hitchcock worked together for a fledgling airline in Western Australia. Like most Australians, they had followed the courageous flight which took four countrymen on the first journey by air from England to Australia. They watched Keith and Ross Smith (no relation to Kingsford Smith) and their mechanics, earn fame and fortune through this spectacular exploit. Very likely it inspired the dream of Anderson and Kingsford Smith to make history by being the first aviators to conquer the Pacific Ocean. Smithy eventually made his dream a reality, and in doing so he reaped enormous rewards and international fame. Anderson, unhappily, moved from one failure to the next.

A brilliant and determined pilot, Anderson was fearless, stubborn, dedicated and hard-working. He also possessed that indefinable trait which separates losers from winners. Fate seemed to keep permanent watch on his movements, making sure that his decisions proved wrong and his best endeavours were frustrated by bad luck.

In short, Keith Anderson was a born loser.

He had been Smithy's partner in the original plan to tackle the Pacific. As the first step in raising sponsorship money, Anderson, Smithy and Hitchcock made the pioneering flight across Australia with paying passengers. They had hoped to set a trans-Australian record for the trip at the same time, but Anderson developed severe tonsillitis thus preventing the attempt from succeeding.

His misfortunes intensified in Sydney where he and Smithy met the dynamic and enterprising Charles Ulm. Within weeks, Anderson realized that he had been relegated from the position of Kingsford Smith's official partner to the junior role in a team of three. This seemed unfair since the object of the team — to conquer the Pacific — was his most cherished dream.

Ulm organized a scheme to attempt to fly around Australia in record time. He hoped that the publicity would attract sponsorship for the Pacific venture. Anderson learned too late that Ulm (who did not have a pilot's licence) would partner Smithy on the record attempt, despite Anderson's extensive experience in the Bristol, which they would be using. Furious, he demanded an explanation. After some angry words, it was agreed that Anderson and

Keith Vincent Anderson, 1898-1929.

Hitchcock would also attempt to create a record, but only after Smithy and Ulm had returned. They did return, after having shattered the record.

Anderson and Hitchcock then took off, hoping to reduce Smithy's record by at least a few hours, but they ran into alarming wind storms. The pair took four days longer than the record which Smithy and Ulm had achieved.

Anderson's confidence and prestige took a further battering, when he found that Smithy and Ulm — on the basis of their successful record attempt — had booked a passage to California, as the first step of their proposed Pacific flight. Anderson's passage had not been booked, and he was shattered. Once again it took a fierce argument before he was reluctantly included in the team.

A new chapter in his life opened when he cemented relations with Smithy and Ulm on the ship carrying them to California. Even then fate continued to be unkind to him. The trio purchased a used Fokker mono-plane which Anderson christened *Southern Cross*. They began to plan the Pacific flight and Anderson was sent by Smithy from San Francisco to Hawaii to investigate possible landing sites for the proposed flight. He returned to find that Smithy and Ulm had organized an attempt on the world record for air endurance, using the *Southern Cross,* to raise essential funds. Anderson naturally assumed that he would be co-pilot, but this was not the case. Smithy had hired an American stand-in.

When the Pacific attempt was almost washed-up in a sea of debts, Anderson concluded that it would never eventuate, so he returned to Australia. As soon as he had departed, luck smiled on Smithy and Ulm. A benefactor — a wealthy ship owner — appeared fortuitously to purchase the heavily mortgaged *Southern Cross,* and he also financed the Pacific flight. Anderson could not believe that such a fairytale could be true, so he declined the invitation to rejoin the Pacific venture.

It is now history that Smithy and Ulm hired two Americans and, together, the four men created an epic chapter in aviation's history — without Anderson's help.

Enormous publicity on both sides of the Pacific followed and Anderson watched Smithy and Ulm reap monumental rewards; the total figure approached $2 million in terms of today's purchasing power. He tried to claim a relatively small amount for himself, but his court action failed.

Determined to achieve his own record, Anderson planned to better Bert Hinkler's time

Bob Hitchcock and Keith Anderson.

on the Australia-England route. Hitchcock would be the co-pilot. However, his Bristol Tourer reached central Australia, where it crashed. The loser syndrome had struck again.

Even minor successes eluded Anderson. Since his plane could not be salvaged, he joined a small airline to fly a Ryan. He was jubilant when his plane managed the fastest time at an air pageant in Queensland, only to find that, after the handicap had been taken into account, the Ryan was placed at fifth position outright.

Immediately after this episode came the unsuccessful lawsuit and Smithy's generous gift of £1,000. Anderson used the money to buy an 80-horsepower Westland Widgeon monoplane (named *Kookaburra*), with which he intended to establish several air records. He was delighted with the plane, especially as it helped

to forge a new bond with Smithy and Ulm. However the bad luck continued. When Smithy decided to continue the Pacific flight and take the *Southern Cross* to England and then the USA — thus being the first aviator to fly the same plane around the world — Anderson again found that he was not part of the crew.

His battle with fate culminated when Smithy and Ulm left Sydney on the first leg of the historic flight. They became lost after experiencing an atrocious storm and finally ran short of fuel in Western Australia. Their disappearance caused the largest air search the nation had seen. Anderson, who was just about to attempt a new endurance record for light aircraft in the *Kookaburra,* decided to join the search.

Lacking the money to properly prepare the plane, he and Bob Hitchcock departed with few supplies, no radio and a faulty compass. After becoming lost several times, Anderson was forced by engine trouble to land off course — in the Tanami Desert. Hitchcock fixed the faulty engine, but Anderson discovered that the plane could not take off in the tangled turpentine scrub and loose sand. They were stranded.

As Anderson and Hitchcock awaited certain death from thirst, Smithy and Ulm were found alive and well. They returned to Sydney, where a section of the Press launched a vicious attack, claiming that the whole episode had been a cheap publicity stunt which had misfired. A subsequent Air Inquiry Committee exonerated the men, but Smithy's reputation was at its lowest ebb. Anderson and Hitchcock, in contrast became public heroes, men who had died for their mates.

However, the story did not end there. The aviators were found on April 21st, 1929; both had died from thirst. The *Kookaburra* was left where it lay. A member of the ground party which subsequently went to the landing site to reclaim the bodies for reinterment in Sydney and Perth, later produced a lump of rock which, he claimed to have found about 30 kilometres from the *Kookaburra*. It was assayed and found to have a high gold content. Rumours spread that the fabulous 'Lasseter's Reef' had been discovered. The rumours triggered a series of searches for the plane and the gold-reef, but the *Kookaburra* could not be relocated. So the alleged gold-reef could not be found either.

It was not until 1978, 49 years later, that aviation enthusiast Dick Smith from Sydney, relocated the skeletal remains of the *Kookaburra* and provided the first convincing theory about the reef.

CHAPTER 2

The dreamers

As they sat sipping beer, Keith Anderson and Charles Kingsford Smith agreed that aviation was progressing at a great rate. But it was not fast enough for them. They wanted to help progress, and that meant achieving something spectacular — and soon.

The year was 1922. Kingsford Smith had just achieved a record time for a mail plane flying between Broome and Port Hedland in Western Australia. Not surprisingly, this minor achievement only managed to gain mention in the local newspapers, but it kindled a burning ambition which both men shared.

Anderson and Kingsford Smith were pilots with Norman Brearley's West Australian Airways. Although five years would pass before Charles Lindberg would make his epic flight across the Atlantic, Andy and Smithy — as they came to be known — were fascinated by the challenge of flying across oceans. Besides, there was a fortune at stake. Three years earlier, Ross and Keith Smith, with mechanics Shiers and Bennett, had collected £10,000 (this is estimated to be worth $400,000 in terms of today's purchasing power), for making the first England-Australia flight. Fame came with the money — and Andy and Smithy wanted both.

Flying mail and passengers over the Kimberley Ranges was fun but it would never gain them a part of aviation's spectacular future. They dreamed of making headlines around the world.

'We could fly from Australia to Africa,' said Andy, pulling on his pipe. 'They reckon it is too dangerous but I've a plan ...'

'No,' replied Smithy, 'let's do the Pacific first.'

The idea was not new. When risking his life while flying as a stunt man for Universal Films in California two years previously, Smithy had talked about being the first aviator to fly from America to his native state of Queensland. In a moment of bravado he had visited a Los Angeles newspaper to tell them of his plans. The editor apparently took the 23-year-old seriously enough to publish his photograph with the caption, 'Captain Charles Kingsford Smith, war ace, who will try to fly the Pacific'.

Anderson's dreams were centred around South Africa because he had been educated there and his friends would surely give him a hero's welcome if he succeeded on the dangerous journey.

A year younger than Smithy, Andy was born in Perth in July 1898. When he was 10 years old, his father accepted a position in Ceylon and young Keith was sent to school in South Africa where he had relatives. When World War I started, he went to England and joined the Royal Flying Corps. With them, he was posted to France with the 73rd Squadron and, at the age of 18 he experienced the intensity of battle. Andy amassed 600 hours of flying experience in nine different types of planes at a time when new pilots were being shot down as fast as they left training school. When the war finally ended, the young Australian was credited with destroying five enemy aircraft and being 'actively responsible' for a further four. He gained the reputation of being the pilot who always came back alive.

Smithy, in contrast, had crashed the fragile collection of fabric, wire and timber in which he attempted his first solo flight. Although he quickly proved to be quite an exceptional pilot, he was shot down and injured when a German Scout dived from the sheltering rays of the sun and caught him unaware. His career as a fighter pilot was over.

Smithy collected a medal for bravery during his limited time at the front and, on his return to Australia, he found himself feted as a war hero. He went back to England and started a flying school, which provided plenty of amusement but not profit. Smithy then sailed to California to commence a new career as a stunt pilot and barn-stormer, but eventually he quit after seeing a fellow pilot crash and burst into flames. He returned to Sydney and joined Digger's Aviation, which had been formed by his war-time comrade, Lionel Lee.

'It will be founded on faith, hope, bunkum and Diggers' war bonds,' he told the grinning Smithy who, at the time, was as irresponsible an aviator as they came. Unfortunately, his disregard for the proprieties of air safety produced more trouble and expensive damage than the young air service could afford. He was given notice, a good flying reference and enough money to take him to Perth. Here Major Norman Brearley (now Sir Norman), had just landed Australia's first air mail contract, flying between Geraldton and Perth, and such isolated communities as Derby, Carnarvon and Broome.

Smithy arrived at the scene just in time to be hired and given the task of test-flying the six

Keith Vincent Anderson.

Anderson's partner, Charles Kingsford Smith.

Bristol Tourers that Brearley had purchased on credit. The planes had been shipped from England and then assembled locally.

In comparison, Andy's early career was very different. He had ended the war with the aviator's equivalent of shell-shock, so returned to South Africa to recuperate. Andy joined the Standard Bank but, like most of his generation, he found it difficult to settle down. He returned to Western Australia where he had been born and accepted a position with Norman Brearley.

It seemed that the new airline would fail, even before it had started. One of the Bristols crashed (killing the pilot and mechanic), only a day after the service had commenced on December 4th, 1921. Fortunately, a subsequent inquiry decided that the Government's failure to provide a suitable landing site (as promised) was the cause

of the crash — and the fledgling airline lived to fly another day.

Brearley's team quickly established an incredible reliability rate of 97 per cent. The five remaining Bristol Tourers between them covered nearly 800,000 kilometres over a five-year period, before being replaced by the more modern De Havilland DH 50.

Smithy and Anderson revelled in the challenge, uncertainty and adventure of pioneering an airline in the days when passengers were understandably frightened and pilots were regarded as super daredevils. Smithy loved being in a different town every night. A born extrovert who was able to play almost any musical instrument by ear, he was at his best at a gathering with a glass of beer in one hand and his well-travelled banjo in the other. He quickly established himself as the principal guest at the parties held wherever his plane touched down.

Quieter by nature, Andy was content to leave the limelight and wisecracks to Smithy, while he dreamed about his plans to show that oceans could — and would — be conquered by aircraft.

Whilst he often talked to Smithy about these ideas, Andy's closest confidant was the airline's ground engineer, Henry Smith Hitchcock, a nuggety young man who answered more readily to the name of Bob.

His background was rather different. Older than his workmates (Hitchcock had been born in Broken Hill, New South Wales in 1891) he had been apprenticed as a bricklayer, working with his father, Harry. This arrangement was not satisfactory, so he took a temporary position with a hairdresser. Harry Hitchcock took his wife and 13 children wherever he could find work and, at various times, the family lived in Adelaide, Perth, Fremantle and Kalgoorlie. Though he literally bounced from one locality to another, Bob Hitchock was deeply religious and remained a regular church-goer throughout his life.

When Dick Smith was researching this book, he spoke to Hitchcock's sister, Mrs Pretoria Bliss, who recalled that when Bob was a lad he would go straight from work to his father's big workshop and lock himself in. One day he emerged triumphant, with a beautifully made model aircraft. His future was moulded by that handiwork.

Hitchcock was later to make his name as a skilful and resourceful engineer, able to turn up a new part or improvise a repair under the most difficult conditions. Yet he had little if any formal training, apart from a stint of work with Atlas Engineering Company at Fremantle, plus a few odd jobs at Kalgoorlie.

In 1911, Bob Hitchcock married his girl-friend Violet, but by then his parents' marriage had broken up; his mother returned to Perth with the younger children.

With World War I, Bob Hitchcock joined the 28th Battalion and sailed to Europe. He was injured at Gallipoli and sent home with a small war pension. Then he worked for a while at the Ivanhoe Gold Mine, Kalgoorlie and later he moved to Perth to take a position driving trucks for the Army Hope Service. However his marriage stumbled from crisis to crisis, with constant arguments about money, and when Bob joined Norman Brearley in 1922, he virtually left home. He continued to support Violet and their three sons (Bob junior, Les and Harry) with his war pension and spasmodic amounts of extra cash.

During his two years of service with West Australian Airways, Bob Hitchcock grew in stature and confidence. His natural engineering

Henry Smith Hitchcock, known as 'Bob', 1891-1929.

abilities were tested to the fullest but were seldom found to be lacking. He proved to be exactly the kind of man needed by a young airline which was struggling to establish itself.

Looking back on those days, Sir Norman Brearley told Dick Smith that Hitchcock was a 'very reliable engineer, ready to overcome difficulties that frequently faced those who made up a team engaged in pioneering work'. He also added that Anderson was reliable and dedicated, well above average in flying skill but that 'he never thrust himself forward as a newsworthy man'. Brearley's highest praise was reserved for Smithy of whom he said:
'He was a really first class pilot of the type needed for overcoming the hazards that faced us in the 1921 and '22 period. Of course, he had to be "tamed" for our course and he submitted when he realized this. His improvement in general behaviour was a gradual process and this development came slowly but surely.'

Confident, boisterous and fearless, Smithy was the natural leader of the small group of employees. He was reckless enough to try anything and skilful enough to retrieve almost any flying situation. Anderson was more of a dreamer, but quickly recognized that whatever

Smithy did would make news, and that headlines were the key to the sponsorship which would be needed for spectacular flying achievements.

Smithy, like Andy, was totally devoted to flying, with the pioneers' determination to see a job through, with no holds barred. It is preposterous to imagine that an airline pilot, finding himself overbooked, would carry a passenger on the wing, but Smithy did precisely this. Sir Norman recalls the time when one of the Bristol Tourers was delayed at Fitzroy Crossing. He telegraphed a former employee, Len Taplin, asking him to fly a DH 50 to Broome to keep the service moving. Taplin did and, on arrival, he found that Smithy (who was to fly the plane out of Broome) had already booked four passengers, not knowing that Taplin was to fly in the plane as well.

Neither man hesitated. Taplin took a 'seat' on the lower wing, close to the fuselage and clutched the bracing wire for the entire 480-kilometre flight to Port Hedland.

Port Hedland was the scene of many of the airmen's memorable parties. It was also here that Smithy met his first wife, Thelma McKenna, although the marriage did not weather the stormy life of an itinerant aviator.

It was during these days that Smithy and Andy solemnly agreed that they would fly across the Pacific and, in doing so, they would be creating aviation history. First they would need to extend their navigation knowledge and raise a considerable amount of money. They threw themselves into their jobs at the airline with renewed enthusiasm and regularly flew over incredibly inhospitable terrain, untroubled by the risks involved. On one occasion, Smithy was forced down by a severe sandstorm and spent three days with only the aircraft wing to shield him from the blazing sun. Badly dehydrated when rescued, he spent the time in hospital calculating the flying hours, fuel load and average speed required for the Pacific flight attempt.

If nothing else, these calculations confirmed his and Andy's concerns that the whole venture was too expensive to contemplate on their own resources. Andy had a wealthy uncle, reputedly one of Norman Brearley's backers, but he rejected the suggestion that he should help to finance the Pacific flight. Disappointed, Smithy mentioned the idea to a wealthy grazier, Keith Mackay. Mackay's enthusiasm was overwhelming. He was prepared to buy the plane and advance the operating cost, on one condition — Mackay would be one of the crew. They shook hands on the deal.

Mackay was already an amateur pilot and the approaching Pacific flight encouraged him to seek further experience. He chartered a light aircraft and, a few weeks later, he crashed into a creek and was drowned.

The Pacific attempt was called off.

Across Australia

During 1924 West Australian Airways made a tidy profit. The pilots, headed by Smithy, decided that the time had come to ask for better pay and conditions. They sent Major Brearley a letter which ended with the threat that if their demands were not met, 'Like Arabs, we will fold our tents and silently steal away'.

At the time, the pilots were paid £600 a year, which was around three times the average salary. Brearley felt that this amount was high enough. He also regarded the letter as a strike threat, so he called the pilots together and said that, apart from enjoying high pay, the pilots often flew only two or three days a week and for the remainder of the time they could do as they pleased. Angry words bounced around the room, after which three pilots, including Smithy and Andy, left the airline. Two remained with Brearley.

Smithy later denied that the pilots were threatening a strike as, 'he would have no part of such a Bolshevik method'. He wrote to his parents saying that although Brearley had agreed to a pay rise, he wanted Smithy to 'climb down in various ways', which he was not prepared to do.

Frankly, neither he nor Anderson was sorry to leave. They had become bored with routine flying and Norman Brearley's strict discipline. Smithy's marriage was 'on the rocks'. He was unsettled, and he knew that if they did not attempt the Pacific flight soon, someone else would.

Andy and Smithy spread the word that they were about to fly the Pacific. An oil company quickly offered free fuel, but no one produced the necessary financial backing. Smithy wrote to an English airplane company (probably A.V. Roe, makers of Avro), suggesting that they should loan a suitable plane, but nothing came of this.

Whilst they waited, Smithy heard about a garage which was for sale in Carnarvon, on the Western Australian coast. Concluding that it would make a good base from which to run an overland trucking firm, he told Anderson that a profitable business could be built up, hauling livestock, wool clips and station supplies. At the time, such goods were mainly carried by camel-train but Smithy reasoned that station owners would gladly pay for a speedier service by truck. Stock and supplies were often ruined by the slow journey and hot weather; and furthermore, delays in the delivery of the wool clip meant that graziers had to wait longer for their money.

In July 1924, Smithy and Andy bought the Gascoyne Transport Company, as equal partners. They acquired a truck on credit and became the local agents for Buick cars. They worked very long hours and the business prospered. As their reputation for fast and reliable service grew, offers of work came from all directions. They purchased a second truck, then a third and, as the profits started to grow, they talked increasingly of the Pacific flight. They continued to seek sponsors, but with no success.

'As a last resort,' Smithy wrote to his parents, 'we will work this truck show at maximum capacity for two years, when we will be able to afford to buy a machine ourselves...'

However, work was not enough. The business needed organization. Andy and Smithy had little aptitude for paperwork. Their accounting system was almost non-existent; some clients were never billed, others did not bother to pay. The business would have gone bankrupt had not Smithy asked his brother-in-law, Bert Pike, to take over the office and handle the accounts.

It was largely due to Pike's management that Smithy's prediction in his letter to his parents came true, almost on schedule. They reached the point where Gascoyne Transport Company had become a sizable concern, with six trucks and a Government contract. Smithy and Andy decided that they could afford to take a holiday. They agreed to travel around the country and, at the same time, look for a sponsor for the Pacific flight.

First, the pair contacted the Western Australian and Federal Governments, but received only luke-warm responses. They departed for Melbourne and Sydney, calling on major firms to try and interest them in backing the flight, but found little enthusiasm.

If nothing else, the holiday convinced the partners that their only hope lay in financing the venture themselves. As their assets consisted almost entirely of the truck company, they agreed to return to Carnarvon to sell it. Fortunately, Bert Pike had done an excellent job as manager and in November 1926 they sold it for £2,300 which was a considerable sum in those days.

With Charles Kingsford Smith and Charles Ulm, Keith Anderson and Bob Hitchcock flew two Bristol biplanes from Perth to Sydney. It was the first flight to cross Australia with paying passengers.

Even this amount was insufficient to finance an attempt to fly over the Pacific, so they divided the cash equally. Smithy went to Sydney, Anderson to Brisbane. The break did not last long as Andy soon contacted Smithy with good news.

Their old mate, Bob Hitchcock, who was still working for Norman Brearley, had written to say that West Australian Airways were selling their old Bristols and that he, Hitchcock, had a splendid idea. They should buy a couple, as they were going cheaply, and fly to New Guinea.

Hitchcock had heard that the isolated gold-mines in that country were experiencing considerable trouble in transporting supplies to those areas, and returning with the gold back to civilization. An air service could make a fortune and — maybe — they might even stumble on a 'find' of their own. Hitchcock was leaving nothing to chance as he had already secured a miner's licence. All that was required now would be to buy the planes and fly to New Guinea.

Smithy and Anderson became as enthusiastic as Hitchcock. They returned to Perth, saw Norman Brearley and purchased two Bristols. They also appointed Bert Pike as the manager of the New Guinea adventure.

History does not record whether Smithy or Andy thought of the idea first but within days they had agreed to 'kill two birds with one stone'. They would fly to New Guinea via Sydney, thus breaking the Perth-Sydney record en route.

The Bristol Tourer was a three-seater biplane which had been derived from the famous war-time fighter plane. With the capacity to carry the pilot, two passengers and 45 kilograms of mail or cargo, it could cruise at around 150 kilometres per hour. Power came from a 230-horsepower Siddeley Puma six-cylinder water-cooled engine. The wing-span was 12 metres, the length eight metres and the height, three metres. Several versions were built jointly by the British Colonial Aeroplane Company and Bristol Aeroplane Company.

A total of eight Tourers was sold in Australia. One was delivered to Colonel Brinsmead who, as controller of Civil Aviation, flew it for 15,000 kilometres surveying new routes. Possibly impressed by this feat, Brearley placed his order as soon as he had signed his air mail contract. Eventually all eight Tourers crashed, but some had long and colourful careers before their final flights.

The two Bristols purchased by Smithy and Anderson eventually made headlines by flying around the continent. Before that, they had been widely acclaimed for successfully crossing the country, with virtually trouble-free flights.

The decision to fly from Perth to Sydney meant that considerable money would be needed for petrol and other expenses. Smithy solved this neatly by suggesting that they should take paying passengers. With his customary eye for publicity, he made sure that one was Mr J.W. Marshall, the Perth representative for the *Guardian* newspaper in Sydney, who was accompanied by his wife. A third passenger, John Howard, also wished to go. He possibly realized that it would be an historic trip, being the first commercial flight across Australia and probably the longest fare-paying flight on record at that time.

The planes took off on schedule, with Smithy and the Marshalls in one plane, Andy, Hitchcock and Howard in the other.

The previous record for crossing Australia by air — a non-commercial flight — had taken five days, and was set by Lieutenant F.S. Briggs in a biplane powered by a 360-horsepower Rolls Royce engine. It was much faster than the Bristol, and had travelled from Perth to Sydney in 21½ hours of actual flying time. Nevertheless Smithy intended to break this record by flying for longer hours each day. As the Bristol had open windows, the noise and general discomfort must have been considerable, but not enough to daunt Mrs Marshall.

'I didn't think anything of it at the time,' she later told journalist Winfred Bisset. 'My husband rang up one afternoon when I was at a bridge party and said, "I'm flying to Sydney next week, would you like to come?" I said "Yes", and went back to the bridge table.'

Mrs Marshall evidently had misgivings, as the trip meant leaving two young children with her parents. Safety factors were poor by modern standards and crossing the country in a single-engined plane, with canvas superstructure, took considerable courage. She must have felt even worse when she eventually saw the Bristol, because many torn patches had been crudely repaired with large stitches and it looked terrible. She recalls asking Keith Anderson if the flight would be dangerous. He looked her straight in the eye and said, 'Mrs Marshall, my neck is as good as yours'.

That convinced her.

'My husband and I wrote notes to each other all the way to Sydney,' she told Winfred Bisset. The ride must have been uncomfortable as well as noisy, as the plane ran into strong winds which buffeted them across the continent.

'By the time we reached Boulder Racecourse, a landing ground, the newspapers had got hold of the story and we could see a huge crowd on the ground and cars, looking like ants, approaching the racecourse.

'A reception committee was waiting for us. It was most embarrassing. There were speeches, and I had to make one, as they told me I was the first woman to fly across Australia.

'In Parkes (NSW), they gave us the freedom of the town and babies were held up for us to see. In Sydney I had to speak over the air and I even had to appear on the movie news-reels.'

The journey itself was fairly uneventful. Smithy and Anderson, who were both excep-

Anderson, Smithy and Hitchcock watch as one of their passengers on the flight across Australia, J.W. Marshall, steps down to join his wife.

ventures, but all failed through lack of money.

Within minutes of their first meeting, Smithy and Ulm knew that they were kindred spirits, each with a passionate ambition to further the cause of flying. Their instant rapport was almost uncanny. In his life story, Smithy later said that they immediately entered into a 'verbal agreement to attempt the Pacific flight together', (this later became a legal partnership).

Anderson's views of the proposal are not on record, but the new arrangements were not to his satisfaction. Although he remained a full partner in Interstate Flying Services (Ulm was appointed manager), Anderson found that he was soon being elbowed to one side by the able and self-confident newcomer.

Regarding Ulm's proposition, he informed them that he had already submitted a scheme to operate the Adelaide-Perth air mail service, but this had been rejected. Now he wanted to re-apply, because the new partnership not only owned two planes, but also had the flying expertise of Smithy and Anderson. With Ulm in the firm, they tendered for the contract on several mail services, but none was accepted. Ulm — a shrewd and far-sighted businessman — quickly realized that the constant discussions about flying across the Pacific Ocean had considerable merit. If it succeeded, the trip would virtually guarantee a large air mail contract.

Whilst Ulm investigated possible sponsors, the business was kept fairly busy with charter and joy-ride trips. Anderson moved into the home of Mr and Mrs A.V. Hilliard at Mosman, an inner suburb of Sydney. It was not long before he became engaged to their daughter Lyal, a cheerful girl usually known as Bon.

Bob Hitchcock was less comfortably placed. He had settled into a boarding house and was accepting almost any available job, working long hours to support his wife and three sons who lived in Perth. According to his sister, Mrs Pretoria Bliss, Bob's wife was constantly asking for additional money.

From all accounts, Hitchcock was a gem amongst aviation mechanics, who was willing to work all night if necessary to prepare a plane for a flight. He thrived on challenges. When there was something unusual or exciting in the offing, his enthusiasm transcended all thoughts of personal fatigue or reward. He was however chronically broke, partly because he tended to under-sell his undoubted skill and partly because aviators, as a class, were often as short of money as he and therefore did not always pay for his services.

Hitchcock continued to work with Interstate Flying Services on a first-call basis and remained an unofficial member of the 'Pacific team'. The unwritten agreement was that Smithy would plan the flight, Anderson would be relief pilot, Ulm would handle the business side and Hitchcock would be chief mechanic. It was Ulm, the resourceful businessman, who made the others fully aware that the plan was doomed to failure unless they could secure definite commercial backing. This meant that they would have to convince potential sponsors that they could offer plenty of good publicity in return for financial support.

It was Ulm who also suggested that before announcing the Pacific flight they should tackle a less ambitious but thoroughly newsworthy project. What better way was there than to break the existing record for flying around Australia?

CHAPTER 4

Around Australia

By mid-1927 only two successful attempts had been made to circle Australia by air. Wing Commander Goble and Flight Lieutenant MacIntyre flying in a seaplane were the first, taking 43 days. Their time was later halved by Colonel H.C. Brinsmead in a De Havilland DH 50, owned by the Department of Civil Aviation. His pilot was E.J. Jones.

To ensure a successful flight several problems had to be overcome. One was that petrol supplies were not normally available and intrepid aviators would need to lay down additional fuel in case head winds or mechanical problems prevented completion of the day's schedule. Night flying was virtually impossible. Apart from the obvious navigation difficulties, few airfields were suitably equipped. Emergency landings after dark were hazardous, as aircraft were not usually fitted with search lights and no radio beacons were there to help.

The existence of a radio was rare but not unknown. Australia's Wireless Telegraphy Act dated back to 1905 and soon after, the first direct radio transmission from England to Australia was sent by the Marconi Company and picked up in Sydney by Mr (later Sir) Ernest Fisk. The Commonwealth Government then sponsored tests in radio telegraphy between Tasmania and the mainland.

Numerous ground-to-air experiments were later conducted, including some by Qantas in 1925, but regular services did not eventuate until nine years later.

In 1927, when they were planning the around-Australia attempt, Smithy, Ulm and Anderson did not have anywhere near the amount of money required to equip the plane with a radio and to set up beacons around the country along the proposed flight path. The trip only became possible because Ulm organized a deal with a group of interstate newspapers. He offered exclusive reports on their progress, in exchange for money for fuel and other expenses.

Ulm did not tell his backers that the record attempt would be risky and unlikely to succeed. Their Bristol, purchased from West Australian Airways, was seven years old and its engine had amassed over 1,100 running hours — which was equivalent to a flying distance of 160,000 kilometres. The plane's design was obsolete, with an open cockpit, primitive navigation aids and insufficient power to climb above storms. Yet if Smithy were to set a new record of 10 days, he would need to average 1,150 kilometres daily, flying strictly between dawn and sunset. This did not allow for mechanical delays or those caused by weather.

Ulm was even more cautious in what he told Keith Anderson. As manager, Ulm had arranged all the details and the scheme was ready to proceed before Andy discovered that Ulm — not he — would be in the co-pilot's seat. Anderson was furious. Apart from other considerations, Ulm was not even a licensed pilot (although he could fly), nor was he a competent mechanic. Smithy was certainly useful with tools, but he was not in the same class as Hitchcock, with regard to fixing an ailing engine quickly.

Smithy and Ulm with the Bristol Tourer which they flew 12,000 kilometres around Australia in the record time of 10 days and five and a half hours.

Queensland Museum.

A light aircraft greets Smithy and Ulm as they finish their record flight around Australia. One of the many signatures on the photograph is that of the Prime Minister of the time, W.M. Hughes.

After what was probably a blistering row, with Smithy backing Ulm, a compromise deal was proposed. Ulm would fly with Smithy, as arranged, but he would also find another backer for a similar attempt to be made by Anderson and Hitchcock in the firm's second Bristol.

Ulm kept his word. He secured financial backing from Messrs George A. Bond and Company, (the hosiery makers), who had just started to manufacture Sidcott flying suits. Nominally at least, Bond would charter the plane, so that the attempt could be billed as the 'first commercial flight around Australia'.

The Bond Company announced that it had chartered the plane to test the 'commercial possibilities of aviation'. There was however, a certain condition, which Anderson learned about after the deal had been finalized. The plane used by Anderson and Hitchcock would not leave Sydney until Kingsford Smith's plane had either completed or abandoned its proposed mission.

Smithy meanwhile had taken the Bristol which was in the better condition of the two, and set about overhauling it, with help from Sydney engineer Syd Marshall. This fact suggests that strained relations had developed with Bob Hitchcock over the flight. To make it possible to stay in the air for 12 hours at a time, extra fuel tanks were fitted with further provision being made for twelve 18-litre petrol cans. Ulm had the unenviable task of tipping the emergency fuel into the main tank whilst the aircraft was in flight.

Eventually, all was ready. Smithy and Ulm took off from Mascot aerodrome in Sydney, at 10 am on June 18th, 1927. They were near Newcastle in New South Wales, when the engine emitted a cacophony of strange noises.

'About two miles* from Boolaroo,' Ulm cabled the *Mail* newspaper in Brisbane, 'Captain Kingsford Smith told me through the earphone that a forced landing was inevitable. The racecourse was the only possible ground, and whilst this was quite large enough in total area, it became apparent when we were within 300 feet** of it that the surface was very bad. It was covered with large, hard tufts of grass, and scattered with logs and burnt-off timber. Large pot holes more than covered the ground and, to make matters worse, in the adjoining field, over

* *3.2 kilometres*
** *91.45 metres.*

which we had to approach the racecourse, were fairly high trees and some houses.'

Smithy made a safe landing and they learned that there was no hope of repairing the engine on the spot. They decided to take the train to Sydney and prepare the second Bristol for the flight.

If nothing else, this decision shows the degree to which Anderson had been relegated to a minor role in the partnership. He and Hitchcock were working hard to prepare 'their' Bristol for the flight, when Smithy arrived and told them to go to Boolaroo and collect the disabled plane.

How much of a storm this created is not known, but eventually Anderson and the faithful Hitchcock left for Boolaroo. Smithy worked through the night to equip the second plane for an early morning take-off. According to Harold Affleck, who later worked as Smithy's chief aircraftsman, Smithy was working on a plank, high in the air. He was so tired that he stepped off and crashed to the ground below, thus suffering painful bruises.

'He told me later that he was really rattled by the fall and "damned near didn't go"', says Affleck. But Smithy forced himself, and just as in later life, he allowed neither illness nor fatigue to hinder a record attempt.

He and Ulm left Mascot at 9.55 am next day, June 19th. They reached Brisbane in six hours and 20 minutes and, as a result of newspaper publicity, the men were greeted by a huge crowd at Eagle Farm aerodrome.

The first minor disruption occurred the next day, on the way to Longreach. A gust of wind caught Ulm's map and blew it out of the window. As this was their only map, Smithy landed at Malvern Hills to ask directions. Despite this delay, they quickly reached Longreach, (then the home of Q.A.N.T.A.S., the Queensland airline which later became Australia's international carrier, Qantas Airways Ltd,). Smithy and Ulm promptly claimed a new record — by flying from Sydney to Longreach in two days.

They set their sights on another record — Australia's longest non-stop flight — which they hoped to achieve on the next hop, by flying direct from Longreach to Darwin, a distance of 1,200 kilometres in 12 hours.

The flight nearly caused the end of the around-Australia attempt.

The engine ran faultlessly until they were within an hour's run of Darwin, when one cylinder faltered. Fortunately, Smithy was able to 'nurse' the engine to Darwin where they found that an exhaust valve had failed. As no spares were available, Smithy and Ulm agreed on the dangerous plan of flying the ailing plane to Broome, in Western Australia, which meant covering the record-breaking distance of 1,374 kilometres. Fortunately fuel supplies did not present a problem, as Broome was on their route and Ulm had previously arranged for petrol supplies to be delivered by camel-train!

The reason for flying direct to Broome was that West Australian Airways had a terminal there and, as former owners of the Bristol, they would almost certainly have engine spares.

To reach Broome direct they would have to fly over desolate country, without the reassuring presence of a radio. If the ill-treated Puma engine were to fail, the plane would be forced down and the chances of survival in that terrain were slim. To make matters worse, the route took them over the Kimberley Ranges, a mountain barrier which Smithy described as 'bristling with buttes and rocks and roughly timbered slopes'. He had flown over it before and he knew what could happen if they were to run into heavy weather with an ailing engine. He knew too that even if they were fortunate enough to find an emergency landing site, they would have no way of sending out a distress message. It could be weeks before they would be found, if ever.

Despite the risks, Ulm and Smithy pressed on cheerfully. Astonishingly, the engine did not fail, the weather remained calm and they reached Broome safely. Although they were tired, the two men had no alternative but to strip down the engine, repair the damage and check anything else which would be likely to malfunction. They hired one of Norman Brearley's mechanics, but the task still took 12 hours.

With scarcely any sleep, the pair took off for Carnarvon at dawn, and ran into winds and heavy rain, which forced them to huddle in the cockpit to keep warm. The light failed before they reached their destination, but Smithy was flying over familiar terrain so he was able to land at Minilya station. The next morning saw an early start for Carnarvon and a touch-down on a landing strip so drenched with rain that it would normally have been closed. After a brief stop to refuel, they left for Perth.

Once again heavy rain tried to wash them out of the sky. At one point the cloud was so thick that they flew out to sea. Fortunately, Smithy had an incredible sense of direction and was able to find the coastline again. He then flew at tree-top level, following the telegraph lines which he knew would lead to Geraldton. From there they were able to fly on a straight route to Perth.

Ulm would later write that it had seemed impossible that they could reach Perth that night. However Smithy was determined to press on and they had landed at Perth in near darkness.

At this point they had been flying in rain for 2,200 kilometres, so they were soaked through to the skin more times than they could count. Thereafter the sun began to smile, and they slowly grew confident of success. They flew across the Nullarbor, made a refuelling stop at Adelaide and finally landed at Melbourne. Here, after warming drinks and a good meal, they sat up into the 'small hours', discussing the Pacific flight.

Next day the Bristol landed at Sydney. Ulm's telegraphed accounts of their exploits had been widely read and there was so much interest in the flight that the Premier of New South Wales, Mr Jack Lang, headed the enormous crowd that greeted the conquering heroes.

Even more newsworthy was the presence in the official enclosure of Nellie Stewart, the famous actress. She placed laurel circlets around the aviators' necks and heartily kissed them for the news-reel cameras. The men had flown 12,000 kilometres around Australia in 10 days and five and a half hours.

Ulm's newspaper publicity and Smithy's jaunty confidence attracted Sydney's leading businessmen to a welcoming party. There they were told of the proposed Pacific flight. Smithy pleaded for financial support and, next day, the

Premier announced that the Government would guarantee the sum of £3,500 towards the cost. Smithy and Ulm booked their passages to the USA.

Meanwhile, Anderson and Hitchcock were unaware of these events. As soon as Smithy had returned to Sydney, they rushed the final preparations for their 'commercial flight'. Anderson was determined to beat Smithy's new record by as large a margin as he could.

Part of the deal with George A. Bond & Company was that they should take one of its staff as an observer. The man selected, the company's display manager Charles Vivian, also acted as the official photographer. The courageous Vivian had volunteered for the assignment and subsequently produced an interesting booklet called *Australia From the Air*. It was published in 1927 by Messrs G.A. Bond & Company.

The booklet made the point that although Smithy and Ulm had flown around the continent in 10 days, the Bond-financed trip was undertaken to 'investigate the possibilities of commercial aviation in Australia', and that it had taken 14 days, 'after having devoted sufficient time to observation, so as to make the expedition of some practical value to aviation'. Undoubtedly this comment was made to assuage Keith Anderson. Anyone reading the

Kingsford Smith and Charles Ulm are congratulated after breaking the record for flying around Australia in the Bristol Tourer.

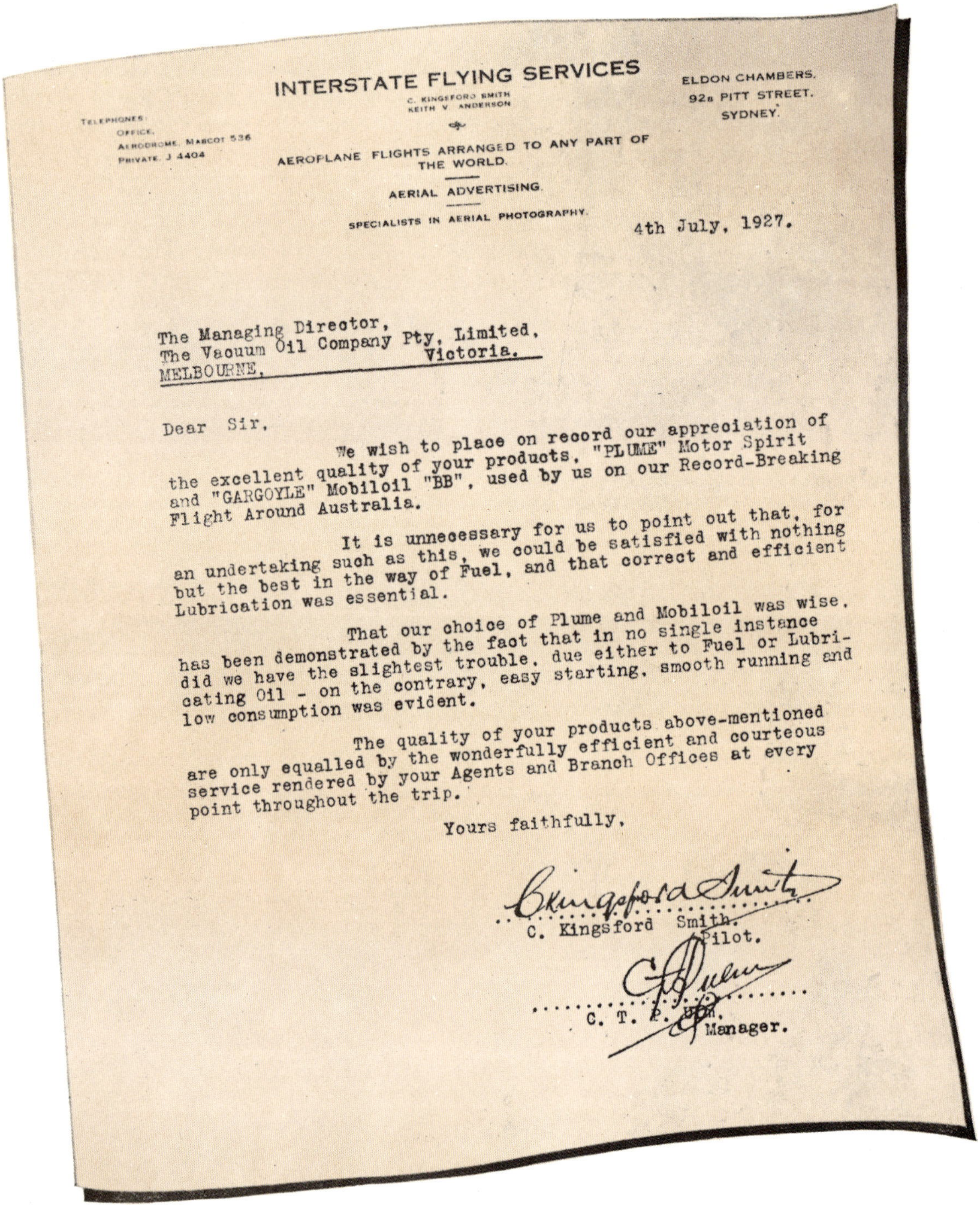

INTERSTATE FLYING SERVICES

C. KINGSFORD SMITH
KEITH V. ANDERSON

ELDON CHAMBERS,
92a PITT STREET,
SYDNEY.

TELEPHONES:
OFFICE,
AERODROME, MASCOT 536
PRIVATE, J 4404

AEROPLANE FLIGHTS ARRANGED TO ANY PART OF
THE WORLD.

AERIAL ADVERTISING.

SPECIALISTS IN AERIAL PHOTOGRAPHY.

4th July, 1927.

The Managing Director,
The Vacuum Oil Company Pty, Limited,
MELBOURNE,   Victoria.

Dear Sir,

We wish to place on record our appreciation of the excellent quality of your products, "PLUME" Motor Spirit and "GARGOYLE" Mobiloil "BB", used by us on our Record-Breaking Flight Around Australia.

It is unnecessary for us to point out that, for an undertaking such as this, we could be satisfied with nothing but the best in the way of Fuel, and that correct and efficient Lubrication was essential.

That our choice of Plume and Mobiloil was wise, has been demonstrated by the fact that in no single instance did we have the slightest trouble, due either to Fuel or Lubricating Oil - on the contrary, easy starting, smooth running and low consumption was evident.

The quality of your products above-mentioned are only equalled by the wonderfully efficient and courteous service rendered by your Agents and Branch Offices at every point throughout the trip.

Yours faithfully,

C. Kingsford Smith.
Pilot.

C. T. P. Ulm.
Manager.

The origin is unknown, but this photograph was probably taken at the start of Anderson's attempt to beat the around-Australia record flying time. Bob Hitchcock pretends to be shy as his colleagues take their farewells. The other aviators are probably Keith Anderson and Charles Vivian.

booklet — even without knowledge of the rift between Anderson and Smithy at the time — would certainly have realized that it was an air race from the moment that the second Bristol left Mascot aerodrome.

To ensure that maximum fuel supplies could be carried, the men loaded very little luggage. Vivian carried a new lightweight movie camera and a celluloid toothbrush, 'as I firmly believe this would be lighter than the ones we are accustomed to use'. As though to explain that they were three men in a boat together, he told his readers that from henceforth the pilot would be referred to as Andy and that H.S. Hitchcock 'really answers only to the name of Bob'.

It is interesting to note that Interstate Flying Service's official letterhead showed Kingsford Smith and Anderson as the principals (at the top of the letter), while Ulm signed the letter as manager. Ulm had effectively become the dominant partner by that time.

The little biplane cruised at an altitude of around 915 metres, climbed to 1,630 metres to pass over mountain ranges, and reached Brisbane after five hours and 40 minutes. Already they were 40 minutes ahead of Smithy's time.

Heavy weather en route to Longreach next day, decreased this advantage and Vivian noted that the ride was so bumpy that, at times, his head was in collision with the cabin roof. He noted the strained expression on Andy's face, adding that telephone communication had been fitted between the pilot and the cabin occupied by Vivian and Hitchcock.

'Scarcely a word was ever spoken during one of these long flights,' he wrote. 'Many times I looked through the tiny window in the cabin directly in front, only to see the intense expression on pilot Anderson's face as he would negotiate difficult passages of air currents. Every time we were tossed to one side by one of these, his unerring judgement would just as quickly correct our angle. The strain must have been great, and his only recreation on landing was a quiet smoke alongside the plane while it was being cared for by Bob.'

At Longreach, they visited the headquarters of Qantas. The airline was described as 'the most up-to-date and complete aerial mail and passenger service in the Commonwealth',

Bob Hitchcock (left), Charles Vivian and a youthful Keith Anderson pose in front of their Bristol during their trip around Australia in 1927. Vivian was virtually a guest, since his firm, George A. Bond & Company, was sponsoring the flight.

adding that Qantas planes could travel from Longreach to Cloncurry in four and a half hours, compared to three days by train.

After the plane had left Cloncurry, Anderson flew in the wrong direction. Instead of heading north-west to Camooweal, he pointed south-west. When it became necessary to land at Urandangi to pinpoint their location, they found that they were 160 kilometres off course.

Vivian did not explain the error, which was extraordinary for a pilot with Anderson's experience, but added that they were soon in the air again 'racing the light', to make as much distance as possible before dark. 'The last ray of light had almost gone,' he said, 'when, picking a suitable paddock, we made a splendid landing.'

They ate their emergency rations, slept in their flying suits and reached Camooweal early next morning where a 'tired individual greeted us'. They then flew over the barren plains of the Northern Territory to Newcastle Waters, slept in the plane overnight and set out for Darwin.

The people of Darwin turned out en mass to greet the aviators and treat them to lunch. At this stage, Andy's time was behind that of Smithy's record, so he decided to omit Katherine from the route and fly straight to Wyndham. Flying conditions were perfect but, unknown to Andy, the town's folk had prepared a special landing strip for them just on the outskirts of the main settlement and a sizeable reception committee had assembled. Having no radio communication, Andy did not know this so he chose another place to touch down, at 5 pm. He wondered why the town seemed so deserted.

Before leaving Sydney, Andy had been warned not to fly directly from Wyndham to Derby because the mountainous terrain would make any forced landing dangerous. The recognized route was to fly to Halls Creek, then follow to telegraph wires to Derby, thereby adding an extra 250 kilometres to the route. Andy — clearly with the record in sight — disregarded this advice and announced that he would fly over the Kimberley Ranges direct to Derby.

The plane was 160 kilometres from their destination when they were forced to climb to the maximum height (3,648 metres), to pass over the 'saw-like edge of the mountain, with its deep gullys and towering cliffs'. Vivian did not say how he, (who was supposedly investigating the commercial possibilities of aviation), felt about risking his life in a single-engined plane over such terrain, but his writings make it clear that he was worried. Fortunately, they reached Derby safely.

Despite this time-saving effort, the Bristol's time was still behind that of Smithy's record, so Andy lost little time in taking off for Broome. There he said he would try to reach Carnarvon in one day, which would necessitate the earliest possible start, with a refuelling stop at Port Hedland. Although he was flying over familiar territory, Andy could do nothing about the blustery weather conditions.

'Bob and I registered several nasty contacts with the top of our heads and the cabin roof,' wrote Vivian. 'Bob's expression showed his displeasure for he huddled in a corner, dug his hands deep in the pockets of his flying suit, and gazed out on the rugged mountainous country and undulating hills that seemed an age passing beneath us. We were scarcely going more than 85 miles per hour*.'

The deteriorating weather conditions provided a taste of what was to come. When they left Carnarvon the next day, they met heavy clouds and, after refuelling at Geraldton, they ran into drenching rain and turbulent air which tossed the plane around the sky as if it was a little bird. Even when Andy climbed to maximum height the rain was almost horizontal, driven by terrific winds. The engine was as unhappy as the aviators and almost stopped on several occasions.

The rough terrain below offered little hope of a safe landing, so Andy turned east and finally flew around the storm. Although he inevitably became lost, he knew the terrain well and was soon able to pick up some land marks. Having done so, he pointed in the direction of Perth where upon their arrival, they were greeted by a huge crowd which had gathered to see the plane. The aviators were advised that a reception had been arranged for the following day — but Anderson shook his head. All pretence of a 'commercial flight' had been dropped. He wanted to make an early start in the morning, for he was determined to beat Smithy's around-Australia record.

Andy's mother was living in Perth at the time. Likewise Hitchcock's mother, as well as his wife and children were there, so time was found for brief family reunions before the trio's departure for Kalgoorlie. It was around this time that Andy read in a newspaper that Smithy and Ulm had booked a sea passage to California on the first step to attempt the Pacific flight. There was no mention of Anderson or Hitchcock in the clipping. An angry exchange of telegrams

* *135 kilometres per hour.*

followed and finally it was agreed that a sea passage would also be booked for Anderson.

Hitchcock felt at home in Kalgoorlie (which they reached in the record time of four hours and five minutes), but the other two found the atmosphere depressing. However there was little time to contemplate the situation as they made a very early start for Cook, which represented the half-way point for the great trans-Australia railway. Trailing winds helped them average 168 kilometres per hour over the Nullarbor Plain, and they reached Cook at 4 pm, giving Bob Hitchcock ample time to service the plane before dark.

Even fresher tail winds helped to push the average speed to 176 kilometres per hour. They touched down at Wirraminna and there they were entertained by a farmer in true bush style — the table was set with enamel plates, large tin mugs, a newspaper table-cloth and tin forks — under the roof of a scrupulously clean galvanized iron shed.

As a result of the helpful winds, they made excellent time to Adelaide, but even so they were still well behind Smithy's time. Any disappointment was at least tempered by the heroes' welcome with which the Adelaide crowd greeted them.

When they flew out next morning, heavy rain and strong winds forced Andy to climb upwards in great circles like an eagle, trying to rise over the heavy white clouds. The rough weather continued into Victoria but, at Melbourne, they were warmed by the throngs of cheering spectators and an official banquet. Their jubilation was marred only by forecasts that further bad weather lay ahead.

Strong winds greeted them again the next morning and, although Andy climbed to maximum height, the turbulent masses of black clouds aided by furious winds, forced him to retreat.

In the teeth of a gale, Andy made an extraordinary decision. According to Vivian's account, he decided to cut his engine and land without power, despite the furious winds. Vivian says that the plane was thrown around the sky and even when they came within a few hundred feet of the ground, it was still being mercilessly buffeted by the winds. At one time they were only a few feet above the tree-tops, 'as Andy's keen eyes sought a safe landing spot'. Apparently he was unable to find one, so he fired the engine again and returned to Melbourne. There they found that the plane which carried mail to various towns in Victoria, had also been driven back by the weather — the first time it had failed to maintain the schedule.

No explanation was given for Andy's highly unorthodox manoeuvre. A pilot would not normally choose to handle an emergency

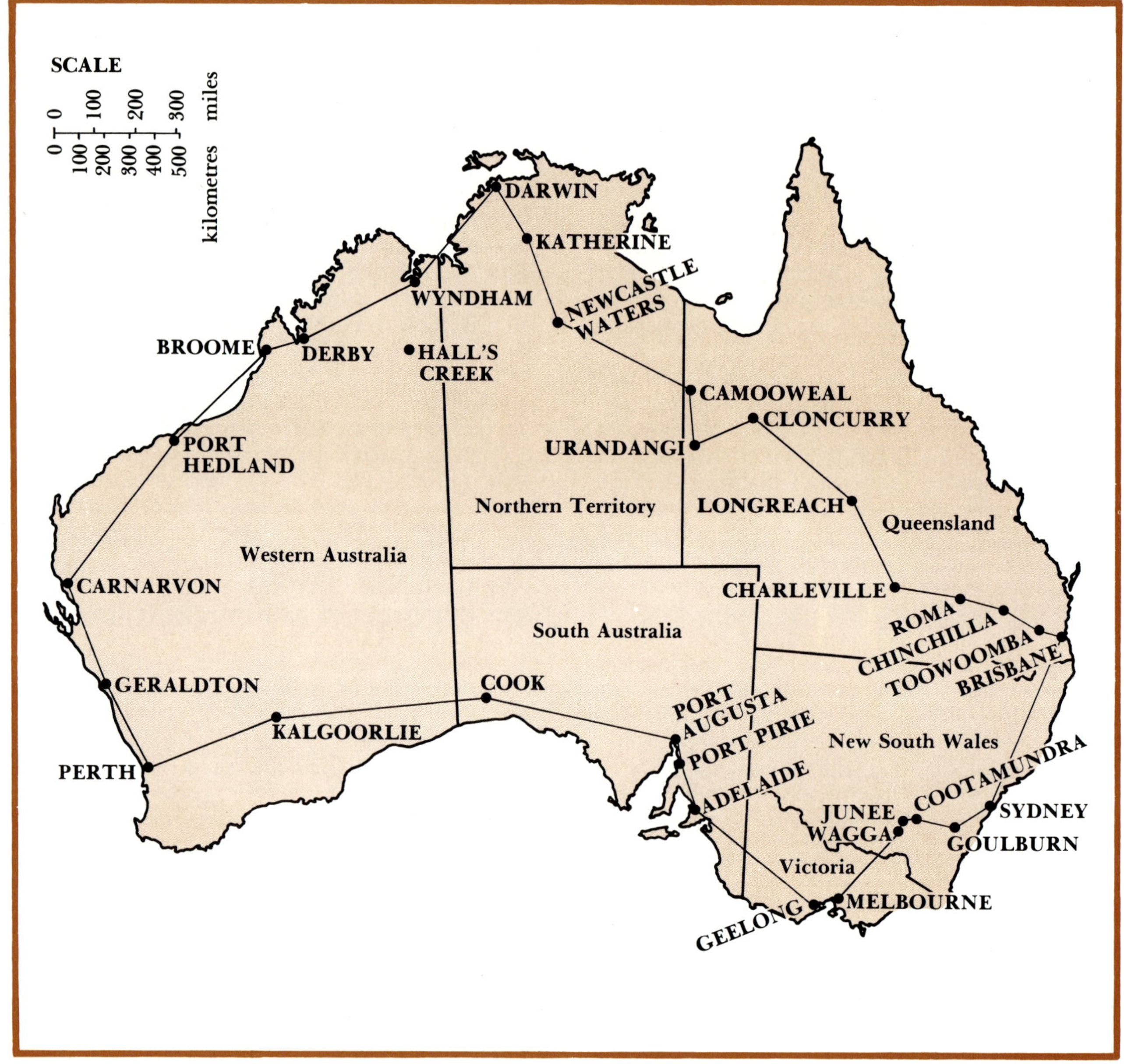

In June 1927, Anderson and Hitchcock set out to break a new record time for flying around Australia, which had been established a matter of days before by Anderson's partners, Kingsford Smith and Charles Ulm. They had an exciting flight but took four days longer than Smithy. The map is taken from a booklet *Australia from the Air* published in 1927 by Charles Vivian who accompanied Anderson and Hitchcock on their flight.

landing without power, especially when forced down by bad weather. Yet Vivian wrote:
'No sooner were we through one rain cloud than we were plunging again into another. Suddenly, during these thrilling moments, I felt our engine cut out ... I afterwards learned that Andy had deliberately stopped it, with a view to climbing down spiral fashion, and this knowledge was a great relief ... as he afterwards explained to me, to have pushed on for a further five minutes would have certainly sent us to our deaths.'

By this time, it was evident to Andy that he had no chance of beating Smithy's record. They waited in Melbourne for a further 24 hours, then took off in less boisterous conditions. The elements decided to have one last tilt at the Bristol before allowing it to reach Sydney. As Andy approached Goulburn, he encountered a hail storm, accompanied by 'terrific winds' and once again he was forced to make an emergency landing. Later in the day they took off again, flew through a snow storm and finally sighted a small flock of Tiger Moths which had flown from Sydney to greet them. The Moths escorted the Bristol to Mascot aerodrome where a civic welcome and an official marquee awaited them.

'Tables were spread,' said Vivian, 'speeches were made by various representatives of com-

At the end of a successful flight around Australia, Hitchcock reads a telegram of congratulations, whilst Vivian smiles for the camera. Kingsford Smith and Charles Ulm can be seen behind Vivian.

mercial and State Government interests.' Afterwards, Andy, Bob and Charles Vivian were driven to Government House to meet His Excellency and Lady de Clair.

They had travelled a total distance of 12,547 kilometres, taking 14 days, of which 87½ hours had been spent in the air. The average air speed was calculated at 90.2 miles (145 kilometres) per hour. The average fuel consumption was eight gallons (36.4 litres) per hour and the average oil consumption five pints (2.8 litres) per hour. The engine used only four pints (2.3 litres) of water during the entire journey.

Despite the speeches and the compliments he received, Andy was deeply disappointed. He knew that Smithy had flown the same distance in a similar plane taking four days less time.

He also suspected from the exchange of telegrams, that Ulm was now clearly established as Smithy's partner for the Pacific flight and that he, Anderson, would once again play the junior role.

Crossing the Pacific

Two weeks after the completion of his around-Australia flight, Anderson boarded the *SS Tahiti* with Smithy and Ulm. A new, friendlier relationship was soon established, possibly because Ulm had become wary of appearing to be the dominant partner. He had negotiated deals with the Sydney *Sun* and Melbourne *Herald* to keep them fully informed of progress. In return, the partners had a firm guarantee of £500. Announcing it, the Sydney *Sun* said that the money would be paid, 'even if the plane does not get past the first stopping place, and much more if it does'.

Ulm faithfully sent regular cables to the newspapers, but he was careful to refer to Smithy and Anderson as the pilots and himself as the navigator. In some cables he made no mention of himself and simply referred to decisions taken by Captain Kingsford Smith and Lieutenant Anderson.

This uncharacteristic modesty probably contributed greatly towards restoring Andy's equanimity.

The ship left Sydney in July 1927, arriving in San Francisco on August 5th. The three men had taken navigation lessons from one of *SS Tahiti's* officers, William Todd, for they were acutely aware that the key to their survival would be the ability to find Hawaii and Suva in the middle of the vast ocean. Failure to do so, meant that the plane would run out of fuel and lead them to a watery grave.

At that stage, Smithy intended to buy a Ryan mono-plane, similar to the one which had just successfully carried Charles Lindberg across the Atlantic. His plan was to fly to Hawaii where floats would be fitted, then set out for Australia in a series of hops, refuelling at Suva and possibly other islands as well.

When they reached San Francisco, the men found the town alive with aviation fever, as the Dole Air Race — from San Francisco to Honolulu — was about to start. Smithy wanted to enter so he examined a single-engined International biplane which was offered to him on loan by an oil company. The Australians thought about their lack of navigation experience and the fact that their plane would be greatly overladen with fuel if it were to reach Honolulu. They reluctantly declined the loan. As it happened, the Dole Air Race (sponsored by a man known as the Pineapple King), was a disaster. Seven lives and two planes were lost. If nothing else, it proved that the single-engined Ryan was too risky to use on the Pacific flight.

Smithy studied various successful long-distance flights and noted that Fokker planes and Wright engines were used for many of them. He concluded that they would need a large Fokker (to carry a substantial fuel load) and that a minimum of three engines would allow them to stay in the air in the event of one failing. By coincidence, the very plane he needed was in California, and owned by notable Australian explorer Sir Hubert Wilkins.

Wilkins owned two Fokkers, one with three engines, the other with one engine. He had flown extensively over Alaska and the adjoining Arctic region and, after the expedition, the planes were dismantled and shipped to the Douglas Aircraft Company. When Wilkins heard about Smithy's proposed Pacific flight, he cabled saying he would be willing to sell one plane, but without the engines or instruments.

The sum involved was £3,000, which consumed almost all of the trio's available capital. Wilkins agreed to accept half the money down, with the rest to follow. Douglas assembled the plane, without the engines and instruments. Smithy explained this to a visiting Australian, Sidney Myer of Myer Emporium. Myer generously wrote a cheque for £1,500, but later changed his mind. He begged the men to call off the dangerous trip, assuring them that they could keep the money for their return trip to Australia. When they refused, he reluctantly gave his blessing and three new Wright J-5 engines were fitted.

The Fokker was test-flown with the word *FOKKER* painted on its wings and fuselage, but Keith Anderson suggested that the plane should be christened *Southern Cross*. This was duly done during a small ceremony, with Mrs Sidney Myer doing the honours.

As soon as the plane was ready, it became apparent that considerably more money would be needed to extend the fuel tanks' capacity to 4,725 litres, so as to provide a flying range of 6,700 kilometres. This work was done, and with the cost of a radio and instruments, the partnership was left in considerable debt.

Meanwhile, Keith Anderson left for Hawaii to inspect the two landing strips — Wheeler Field and Barking Sands.

He returned in September and the men planned to attempt the Pacific flight in October.

However it became apparent that Smithy needed considerably more practice at flying the plane with its huge fuel load and Ulm required further experience in navigating, especially in cloud, as there would be no markers to guide them over the ocean.

The fuel and other costs rose sharply and the partners were soon hopelessly in debt. Their families urged them to abandon the idea and, to make matters worse, the money promised by the New South Wales Government had not arrived. The Federal Government turned a deaf ear to their pleas for help and Sir Hubert Wilkins began applying gentle pressure for the balance of the purchase price of the plane. The trio estimated their total debts at this stage to be £8,000, which would be equivalent to a staggering $320,000 in today's purchasing terms.

A new problem arose. The newly elected Premier of New South Wales, Mr Bavin, announced that the flight would be too risky. He urged Smithy to cancel it and reinforced the message by stating that the money promised by his predecessor would not be paid, because the proposed flight was too dangerous.

The energetic Charles Ulm rapidly visited potential sponsors and came back with financial offers from two oil companies — Vacuum and Fargool. In return, Smithy would make an attempt on the existing world record for sustained flight. This was held by a German-built Junkers plane which had remained aloft over Leipzig for 52 hours and 22 minutes.

The *Southern Cross* was stripped of every unnecessary item, allowing the maximum fuel load to be carried. Meanwhile Keith Anderson had returned from Hawaii, expecting to be co-pilot for the endurance attempt. He was mistaken. Smithy had already hired Lieutenant G.P. Pond, whom he described as 'a very experienced Fokker pilot'.

Smithy and Pond made five unsuccessful attempts to beat the record. They finally loaded so much fuel into the plane that it barely staggered off the ground. It stayed in the air for 50 hours and seven minutes, before landing without a drop of fuel. It was a valiant attempt, but Smithy was well aware that there are no prizes for coming second.

The men were bitterly disappointed and Keith Anderson decided to quit. Some reports say that he had received a cable from Bon, his fiancée, urging him to abandon the idea, but this was not so. Whatever the reason, a dispirited Anderson arrived back in Sydney in March 1928.

The Premier of New South Wales was now insisting that Smithy and Ulm abandon the Pacific flight. They agreed. The financial pressures were intolerable and could only be eased by selling the *Southern Cross*. They flew to Los Angeles, believing that the Vacuum Oil Company would buy the plane and engage them as pilots, but they left disappointed and close to desperation. The sequence of events which followed is like the plot of a Hollywood film-script.

The Australians were introduced to a wealthy shipping man, Captain G. Allan Hancock, a master mariner with an abiding interest in navigation. Smithy explained the proposed Pacific flight and the reason why they had cancelled it. Some days later he and Ulm received an unexpected invitation to join Captain Hancock on a two-week cruise aboard his steam yacht, *Oaxaca*. During the trip, Hancock made an extraordinary offer. He would buy the *Southern Cross* for £8,000, which would be enough to pay off their debts, and then loan the money to them for the Pacific flight. He also suggested that a film should be made of the attempt. He offered to advance additional money for this film, the proceeds of which would finance the flight itself.

Dazed with gratitude, Smithy and Ulm left the *Oaxaca* to find that their plane had been seized on behalf of creditors. The money from Captain Hancock quickly solved this problem.

As Ulm planned details of the flight, Smithy took the plane back to Douglas in Seattle for modification and then to San Francisco where the last of the instruments and radio components were fitted. Ulm sent a cable to Sydney inviting Anderson to rejoin the crew, but he declined. Smithy asked around and eventually met an experienced sea-going navigator, Captain Harry Lyon, of Maine. He in turn knew a first-class radio operator, James Warner, of Kansas City, who also joined the flight.

Smithy was not taking any chances. He flew the *Southern Cross* on a long trip, with Captain Hancock sitting in the co-pilot's seat; Ulm was in the navigator's cabin, where he could watch how Lyon and Warner 'reacted to air conditions'.

The biggest flight problem by far was the question of noise. The thunder from the three engines was deafening to those in the semi-open cockpit. Even in the closed navigator's cabin at the rear, it was impossible to talk. The men experimented with an electric telephone system, but found that they still could not hear, even when messages were shouted. They resorted to exchanging written notes.

By now Smithy was displaying signs of the meticulous care that would become one of his outstanding qualities. For hours he practised flying blind, using the bank-and-turn indicator, in conjunction with the rate-of-climb metre. He learned how to keep an accurate track of his aircraft, holding a steady course for as long as was necessary.

As Smithy planned to carry 5,870 litres of fuel — equivalent in weight to 57 adult passengers — he also had to learn the best method to effect take-off of the grossly overladen plane and control it in bumpy conditions.

On May 31st, 1928, after spending 10 months in the USA, Smithy and Ulm were ready to leave. The Golden Gate Bridge was shrouded in mist when the *Southern Cross* took off from Oakland and flew across San Francisco Bay. It rose slowly into the air, climbed to 300 metres and banked over San Francisco. The flight plan consisted of three long hops, with fuel and rest stops at Honolulu and Suva, and a touch-down at Brisbane, Australia. They expected rough weather. The Pacific sky was notorious for electric storms and, for airmen at least, south of Honolulu the ocean was uncharted territory.

Flying at a speed of 130 kilometres per hour at an altitude of 580 metres, the *Southern Cross* made excellent headway. During the first few hours, it was guided by a radio beam directed over the ocean from the Californian coast. Smithy and Ulm alternated at the controls, with Smithy doing the greater share and all of the blind flying. Ulm relieved him at intervals, kept the log and communicated with Harry Lyon and Jim Warner in the rear cabin.

Conditions were most uncomfortable. The vibration was enough to rattle their teeth and the noise was so bad that it deafened them. Cold air tore into the open cockpit and numbed their fingers. Smithy and Ulm were soaked each time it rained, although Lyon and Warner were slightly better off, since their cabin was closed in and they had enough space to move around. Lyon found it difficult to take sextant sightings, but at least there was an opening in the cabin floor through which he could take drift sightings when the ocean was visible.

The men soon found relief by passing humorous notes, often coloured with spicy adjectives and comments; Harry Lyon had a special gift in this regard.

Four hours out of San Francisco, Smithy climbed to 800 metres. Suddenly the white carpet below disappeared and the plane was engulfed in cloud. He climbed higher but still he could not rise above the cloud. Condition became increasingly bumpy and, for over an hour, they were flying blind. After eight hours of flight, Smithy and Ulm became concerned about the rate of fuel consumption. The primitive fuel gauges (consisting of glass tubes), showed that more fuel was being used than they had estimated. Smithy became so anxious that he decided to run the wing fuel tanks until they were dry, as this would then give an accurate estimate of how fast the fuel was being burned.

When darkness came, they climbed to 1,200 metres to increase the safety margin. Three hours after they had started to check the rate of fuel consumption, the wing tanks drained, reassuring Smithy that it was not excessive. As the *Southern Cross* roared through the night, its bright-blue exhaust flames lit the sky. Occasionally Lyon threw out watertight flares to check the drift.

The night was wet and bumpy.

Morning light revealed great valleys of cloud plunging below them. Smithy and Ulm marvelled at the sight, which resembled a sculptured landscape of mountains and canyons, falling away to white foothills and carpeted plains. The sheer beauty of the view and the colourful horizon, fired by the rising sun, renewed their spirits. Lyon took his bearings, Ulm checked the fuel and a flurry of good-humoured notes was exchanged.

For a short time Lyon lost their position, but eventually Warner caught the Hawaiian radio beacon. Later in the morning they sighted the volcano Mauna Loa, its active twin craters reaching up to the sky. Soon, welcoming planes arrived and escorted them to Wheeler Field, which was 35 kilometres from Honolulu.

After 27½ hours in the air, the *Southern Cross* touched down. It had travelled 3,385 kilometres and still had enough fuel left for three and a half hours of flight. Within seconds of their arrival, the crew faced a battery of cameras and a series of welcoming speeches. They heard none of them. The engine noise had caused the four men to become totally deaf for several days.

Whilst they slept, army technicians serviced the motors, drained the fuel tanks (to accurately check the consumption), and added enough fuel to take the plane to the neighbouring island of Kauai, where the Barking Sands runway would be long enough to effect a take-off with a full fuel load.

The trip to Suva was the longest of the three legs and its aerial route was still uncharted. Smithy resolved to depart with 5,870 litres of fuel, enough for 41 hours of flight. The risks incurred in being grossly overladen were great, especially if the plane had struck severe turbulence before the excess fuel was used. However the crew discussed the problem and dismissed the danger.

At 5.12 am, the heavy plane lumbered into the air. The weight was so great that it took six minutes to climb the first 100 metres. At this height, they ran into turbulence. Ulm watched tensely as Smithy tested all his skill to stop the plane from rocking excessively. After 20 minutes of flight, the *Southern Cross* had only climbed an additional five metres.

For the moment, navigation did not present a problem. Army operators had flung a radio beacon to guide them over the first 1,000

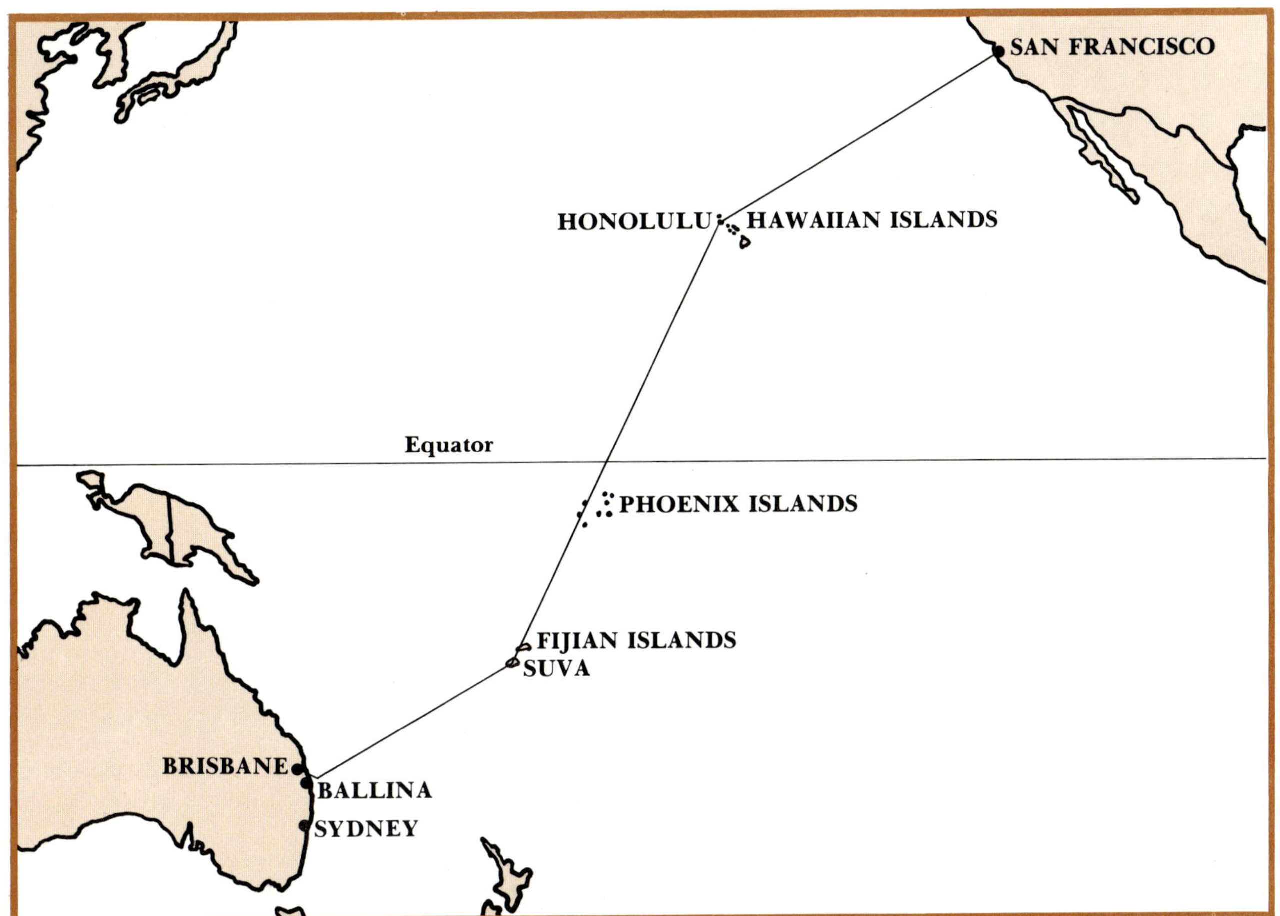

The flight path of the *Southern Cross* from San Francisco to Brisbane.

kilometres of their course. At least, that was the theory. Unfortunately, it was not long before one of the radio receivers broke down due to battery trouble.

Suddenly Ulm spotted drops of liquid from the wing. The drip became a trickle... Was it petrol? If so, would the ignition system spark the fuel and cause an almighty fire? Smithy handed the controls to Ulm and investigated. Thankfully, the liquid proved to be water, condensed from the air by the cold fuel pipes.

Warner passed a note saying that all of the radio equipment had failed. Just as the four men realized that they were isolated from the rest of the world, ominous dark clouds tumbled towards them. Aware that heavy rain storms would increase the consumption of the fuel supply, Smithy decided to conserve it by staying as close to the water as possible. Although he tried to avoid the worst of the storm, the *Southern Cross* was soon engulfed in howling winds and driving rain. Visibility diminished to frightening limits. It grew worse as the storm raged on. Smithy could not continue safely while flying so close to the water, so he pointed the nose upwards. The overladen plane lurched and heaved through great curtains of rain driven by hurricane force winds. Slowly it climbed into the sky.

The storm continued until noon, when conditions eased. Despite the battering it had taken, the *Southern Cross* did not falter and continued to cruise at 145 kilometres per hour. By early afternoon, Warner had managed to repair the faulty radio and Lyon had pin-pointed their position.

Suddenly the starboard engine emitted a tremendous cough — this was the first time it had faltered. Smithy and Ulm exchanged glances of horror. A loud whine developed and Lyon passed an anxious message. Although the coughing continued for some time, it eventually subsided, along with the whine. The cause was never located.

By late afternoon, the rain had ceased, but darkness brought fresh violence. The winds and rain became so heavy that Smithy was forced to fly in circles, steadily gaining height in what he described later as the 'longest and most strenuous climb of the entire journey'. The battle raged on, consuming large amounts of vital fuel. Smithy decided to try and climb above the inky darkness and the plane finally burst through the clouds. Glittering above them was the constellation Southern Cross.

At midnight, they crossed the equator, an achievement that was blunted by the failure of the instrument panel light. Now Smithy was really flying blind. The plane thrashed on through the early hours and the crew abandoned their plans to try and locate the Phoenix Islands for reference purposes. They set course for Suva.

Just before dawn, the plane suddenly dropped sharply, as the wings crashed into an air pocket. This was the start of some appalling weather. Lightning flashed across the skies, illuminating the angry seas below. Wild air currents threw the plane from cloud to cloud. Rain lashed down, as though trying to obliterate the very presence of the intruding plane. Warner and Lyon were forced to abandon any attempt at navigation.

At 7 am, the plane suddenly stopped bucking. Dawn's early light threw an entrancing tint of pink and gold across the sky. Lyon calculated that they were 690 nautical miles (about 1,300 kilometres) from Suva.

The respite was brief. Within an hour they were in the teeth of yet another storm, fighting a head wind which chopped and flung them around the sky. To conserve fuel, Smithy again descended as low as he dared. Conditions deteriorated still further with the trade wind hurling its full weight against the plane. At one point Smithy estimated that they had used 180 litres of fuel without making any progress.

Physically weary and depressed by the seemingly endless battle with the weather, and the dwindling fuel supplies, Smithy was quiet and gloomy. However his spirits rose sharply when Lyon passed a note saying that he and Warner had decided, by resolution of a meeting held in the navigation cabin, that they would have Kingsford Smith elected as President of the USA for the way in which he had ploughed through the arched squalls of the night. Further tonic came when the weather cleared long enough for Lyon to take a sighting. The four were overjoyed to find that they were only slightly east of the direct course to Suva.

Ahead lay an endless plain of water, its waves whipped by the wind into foaming crests. They pressed on scanning the horizon until 1.10 pm, when Ulm sighted a small brown lump peeping out of the sea. Smithy took the plane to only six metres above the water, so that Lyon could use the natural horizon for an accurate sextant reading. He confirmed that they were within an hour's flying time of the Fijian Islands.

At 3.45 pm, after covering 5,265 kilometres in just under 35 hours, the *Southern Cross* banked over Suva and headed for the Albert Park sports oval, where he made a skilful landing in the fenced area.

Next morning Ulm cabled his newspaper reports to Australia, while Smithy set out to inspect some proposed take-off strips. He returned at night with the good news that Naselai Beach, on an island 30 kilometres east of Suva, would be perfect. Smithy then joined the others at a dance which was being held in their honour. Later that night he and Ulm organized the shipping of fuel supplies to Naselai and, when they returned to the dance, the people of the town presented them with 200 gold sovereigns in a tortoise-shell casket.

Suva was bathed in glorious sunshine the next morning as Smithy flew the lightly laden *Southern Cross* from Albert Park to Naselai Beach, where its fuel tanks were filled. On Friday, June 7th, the *Southern Cross* took off to the ringing cheers of the assembled crowd. Their heading — Brisbane.

Shortly after take-off Lyon reported that the earth inductor compass, the most accurate of their three compasses, was not functioning. Using the steering compasses, they flew on through idyllic weather conditions but as the light began to fail, the air took on a heavy and menacing atmosphere. The temperature dropped sharply. Within minutes the plane was bucking violently through a rain storm. Smithy could see nothing beyond the windscreen — not even a propeller. To climb above the storm, he slowly coaxed the heavily laden plane to 2,750 metres but raking winds continued to toss them around the sky. Hissing walls of water blanketed the windscreen and turbulence was so severe that, at times, it took the combined strength of Smithy and Ulm to correct the controls.

Waves of cold air began to slow the motors. Smithy conceived the plan of diving towards the sea because the plane was easier to control at higher speeds. After each dive they would slowly gain enough height to start another one. Ulm's fingers became too numb to continue his log. Lyon passed notes asking what Smithy was trying to do and how long the pummelling would last. Lyon had little else to occupy his mind. For three hours it had been impossible for him to make any attempt at navigation. Smithy just flew instinctively, pointing the plane in the direction which he felt it should go, content to survive the raging storm.

The storm eased after midnight and, although another storm struck in the early hours, the weight of fuel had been sufficiently reduced to allow Smithy to climb above the trouble. As dawn spread across the ocean, Lyon took a sighting. As expected, they had been blown far off course. Fortunately, the whole of Australia's east coast lay ahead, so there was little fear that they would miss it. Jim Warner meanwhile had his radio operating again. He made contact with Australia and told them of the night's ordeal.

From left to right, Harry Lyon, Charles Ulm, Charles Kingsford Smith and Jim Warner in front of the *Southern Cross* after the flight across the Pacific.

'Bully,' said Lyon solemnly as he sighted land. He shook hands with Warner, then checked their position. They were about 176 kilometres south of Brisbane, their target. Smithy recognized the coastal town of Ballina in New South Wales, so he turned north and sped towards Brisbane. The weather was calm now and the jubilant men watched the long ribbon of sand and surf unravel below them.

Smithy touched down to the cheers of 15,000 visitors. Lyon calculated that the 3,000-kilometre trip from Suva had taken 21½ hours, but the average speed had easily been their slowest since San Francisco. During the trans-Pacific flight, the men had spent some 83 hours and 11 minutes in the air. Nearly half of this time had been taken up fighting storms.

After a short stop in Brisbane, Smithy took off for Sydney. The four received a tumultuous welcome from a crowd, unofficially estimated at 300,000, which flocked to Mascot aerodrome. Smithy and Ulm were deeply moved by the welcoming speeches.

As if to acknowledge the historic moment, a cable arrived from Captain Hancock presenting Smithy and Ulm with an unconditional gift of the *Southern Cross*.

CHAPTER 6

The lawsuits

The rewards were enormous. In addition to receiving the *Southern Cross* as an outright gift, Smithy and Ulm were both honoured with the Air Force Cross. Smithy was appointed to the position of an honorary Squadron Leader, Ulm a Flight Lieutenant. They also received a Government cheque for £5,000, a staggering sum equivalent to around $200,000 today. In addition to this, Ulm's newspaper contacts paid off, as the Sydney *Sun* and the Melbourne *Herald* had launched a public appeal which eventually provided a total of £20,000, which is estimated to be about $1 million in today's purchasing terms.

It is hardly surprising that Anderson and Hitchcock felt they should share the rewards. They initiated separate lawsuits. Hitchcock handed his writ to Smithy during a welcome-home dinner. Anderson's suit was conducted by Mr A.V. Hilliard, Bon's father, who had a legal practice in Pitt Street, Sydney.

Hitchcock chose to use a different lawyer, because he had included Anderson in his writ. He claimed that Smithy, Andy and Ulm had verbally promised that he would be the flight engineer during the Pacific crossing, and as such he was entitled to payment.

Surprisingly, Hitchcock and Anderson remained firm friends. Whilst the necessary paperwork was done, the two men decided to make aviation headlines of their own. In September 1928 they attempted to beat Bert Hinkler's record flight between Australia and England, using the old Bristol Tourer Anderson and Hitchcock had flown around Australia.

For some reason, possibly because of their age, the Bristols had not been sold whilst the trio were experiencing their financial difficulties in the USA and, on his return, Anderson regarded one of the machines as his own.

The proposed flight created a controversy, the news of which eventually reached the Prime Minister's office and he called for a report from H.C. Brinsmead, controller of Civil Aviation. He replied that he had interviewed the Attorney General about the legal aspects of the flight and that Anderson had been officially warned of the risks and delays inherent in some countries which he proposed to pass through. Brinsmead added that steps had been taken to check the passports of the two fliers, as Anderson proposed to land at Bima, Surabaya, Singapore, Akyab, Calcutta, Agra, Karachi, Bandar Abbas, Baghdad, Aleppo, Constantinople (now Istanbul), Belgrade, Vienna and London.

Judging by the correspondence, it seems that the Government considered the trip to be inadvisable but could do nothing to prevent it. So, on September 6th, 1928, the ever-cheerful Bon was at Mascot aerodrome to wave the Bristol Tourer off. The plan was to fly from Sydney to London in less than 14 days, thus breaking the existing record. Anderson headed for Queensland, but ran into a wind storm and was forced to turn back. He took off again, reaching Charleville on Sunday, September 9th. The flight to Camooweal passed without incident but radiator trouble did occur on the leg to Darwin. Determined to carry on, Andy stopped at every water-hole between Newcastle Waters and Katherine. He was on his way to Darwin when the radiator overheated again, almost causing the engine to seize. The terrain was quite unsuitable for landing, as Andy could see high timber in all directions. Having realized that a crash landing was inevitable, Andy chose the most suitable spot.

He later said:

'I decided to save the engine as we were in for a crash in any case and where we were at the moment was about the most inviting place to crash from a personal injury point of view. I therefore pancaked the "Old Pioneer" into some trees just short of a little cleared ground about 50 yards* long. We fell pretty heavily to the ground and, through striking other trees, were swung around until we faced the opposite direction from our approach.'

The plane was badly damaged. Hitchcock received a deep cut on his bottom lip, as well as a leg injury and a very bruised shoulder. Anderson was badly shaken but unhurt.

The crash had occurred at Pine Creek and the battered aviators made their way to Katherine. Andy managed to sell the engine to a publican, and the sale provided the money needed to return to Sydney. Possibly pride was speaking when he told a local newspaper that he was returning only to make a fresh start. The Bristol was left where it had crashed but years later another plane met the same fate at Pine Creek and a man by the name of Cox rebuilt its damaged wing using the spars from the old Bristol.

* *46 metres*

Although Anderson reiterated that he would make a second attempt on the record, he was not in a financial position to do so. It is unlikely that the Bristol was insured (his parlous financial position did not run to such luxuries), and the day of his court case was drawing near. He therefore took a position with New South Wales Airways Ltd, where he would be responsible for a 200-horsepower Ryan mono-plane. Named the *City of Sydney,* the Ryan was a fairly versatile aircraft, with seating for five passengers and a cargo load of 27 kilograms.

Similar in design to Lindberg's trans-Atlantic machine, the Ryan was powered by a single Wright Whirlwind J5 engine giving a maximum speed of 200 kilometres per hour. Fairly large for that time, the plane had a wing-span of 12.8 metres, a length of 8.4 metres and a total payload of 364 kilograms.

New South Wales Airways Ltd, intended to establish a regular Sydney-Brisbane service and Anderson volunteered to attempt the record time to attract publicity. Showing a high degree of professionalism, he chose a time which coincided with the first Australia-England test match held at Brisbane's 'Gabba' ground. The *City of Sydney* left Mascot aerodrome at 6.55 am on November 30th, 1928. By noon Anderson and his passenger were seated in the Brisbane Exhibition Grounds for the start of a match which Australia eventually lost. It made an excellent newspaper story.

On Boxing Day of 1928, Anderson featured in the news again. While taking off at Mascot aerodrome, the *City of Sydney* ran into a channel, causing one wheel to snap off and the propeller to shatter. Fortunately, the five people aboard were not injured.

In February 1929 Anderson, supported by Mr Hilliard, appeared in court, to give evidence in the lawsuit against Smithy and Ulm, claiming a share of the profits from the successful Pacific flight. Those profits were considerable, probably far higher than any of the partners had originally anticipated. In direct and indirect payments, including the gift of the *Southern Cross,* Smithy and Ulm had collected in the vicinity of £50,000. To give an indication of that amount's true value, at that time a new house in Sydney could be purchased for around £500.

Anderson's attempt to gain a share of this money was defeated when he admitted in court that his departure from the USA had effectively dissolved the partnership. He also agreed that Smithy and Ulm had cabled, inviting him to rejoin the flight, after they had secured financial support. According to a newspaper report on the verdict, Anderson admitted in court, 'that he had no right, claim, title or interest in any prize money or any other profit of the flight'. Irrespective of Anderson's legal defeat, Bob Hitchcock proceeded with his case four weeks later. On March 20th, 1929 Mr Justice Davidson presided over an action in which Henry Smith Hitchcock claimed £1,000 damages from Kingsford Smith, Charles Ulm and Keith Anderson. Referring to a verbal promise which he claimed had been made, Hitchcock said that whilst working as a ground engineer with West Australian Airways, Kingsford Smith had approached him with an offer to undertake the Pacific flight. He said he had accepted the offer, resigned his job and travelled to Sydney with Smithy and his partner, Keith Anderson. It was allegedly intended that the three men would take the two Bristols to New Guinea to engage in commercial air transport services thus raising money for the Pacific venture.

Hitchcock told the court that in Sydney, the group had made the acquaintance of Charles Ulm who was then without employment. Subsequently Ulm had become engaged by Smithy and Anderson to try and interest various firms in sponsoring an around-Australia flight. Further, Hitchcock maintained that when he had demurred about being left behind, because Ulm was taking the mechanic's seat, he was told that he could use the second plane for a similar trip, to be sponsored by Bonds. Hitchcock added that he had been told that this attempt could not start for a week or 10 days after the flight by Smithy and Ulm.

Hitchcock went on to say that he and Anderson were half-way around Australia when they read in a local paper that Smithy and Ulm, now public figures, had booked their passages to the USA. This meant that Smithy and Ulm would probably leave Sydney before the second Bristol (with Anderson and Hitchcock aboard) had returned to Sydney. Telegrams were exchanged and as a result, said Hitchcock, Anderson had been able to join the party shortly before the day of departure. Hitchcock alleged that he had been told by Kingsford Smith that he must remain in Australia because Ulm insisted on being included on the trip and the plane they proposed to buy was too small to take all of them.

This evidence served as a background to Hitchcock's main case which was that Kingsford Smith had promised to pay him £1,000 as compensation on their return from the USA. Smithy and Ulm replied by assuring the court that no such promise had been made. They had merely said that if the plaintiff looked after the two Bristol Tourers during their absence, 'they might be able to make him a decent present', if the flight was successful.

Although this present never materialized, the court took the view that no contract had been entered into. Hitchcock lost his case.

The Kookaburra is born

An immediate and unexpected consequence of the lawsuits was that Smithy and Ulm decided to jointly give Anderson the sum of £1,000, but there is no record of Hitchcock having received a similar gift. Smithy's generosity could be interpreted as a flash of conscience, but he saw things differently. Commenting on the events in his book, *The Old Bus,* Kingsford Smith wrote: 'I was pained and surprised when, badly advised as I think, he (Anderson) brought a suit against Ulm and myself, claiming a part share in the proceeds of the flight.

'Judgement was given in our favour, but to show our old friend that we entertained only the warmest feelings towards him, Ulm and I voluntarily made him a gift, with which he purchased a Westland Widgeon plane, christened *Kookaburra.*

Smithy's words rather understated the case, for the gift fulfilled Keith Anderson's most cherished dream — to own his own plane. Ever since the loss of the Bristol Tourer, he had been like a captain without a ship. Aviation was his life and livelihood. Flying for other people was uncertain and probably unrewarding, but his own machine would provide his kind of security and freedom.

This fact makes his choice of a Westland Widgeon III rather curious. To draw a parallel it would be similar to a man who desperately needed a light truck, but bought a sports car. But Anderson had a reason, although it was an impractical one. He wanted to record his name in aviation history by creating some world records. Anderson already held a few inter-city Australian records and he had never forgotten his promise to the Press that he would again attempt the Australia-England air record.

The first Widgeon was built for the Lympne air competition of 1924. The Westland Aircraft factory at Yeoville, Somerset in England spent three years on the prototype. Their aim was to produce a simple, high-winged mono-plane without rigging wires, using tubular steel struts for the wing bracing and a semi-monocoque fuselage. For the time period, this concept was advanced and mass production was evidently planned, as the fuselage was built in special jigs, from spruce members and mahogany ply. The original production version was powered by a 60-horsepower Cirrus air-cooled engine, with an Armstrong Siddeley Genet as an option. The latter was described in the sales brochure as 'a small brother of the famous engine used by Sir Alan Cobham in his famous flight from the United Kingdom to Cape and Australia'.

Another feature was that one man could fold the wings in less than a minute and when folded, the little plane was only 3.2 metres high, 2.6 metres wide and 7.1 metres long. It was designed to carry a pilot and one passenger, with a roomy luggage trunk and enough fuel for a four-hour flight at around 136 kilometres per hour. With extra tanks, the range could be extended to a 14-hour flight, according to the manufacturers. Claiming a maximum speed of 160 kilometres (100 miles) per hour, they said that the Widgeon would average '20 miles to a gallon of ordinary petrol obtainable from any garage'.

The design was available for £750 from the factory in England.

In 1926, Westland decided to introduce a modified Widgeon which would be suitable for private owners. Known as the Widgeon III, it was first flown in 1927, powered by an 80-horsepower Cirrus engine. The basic features of the first model were retained, but constant chord wing ribs were used to simplify the construction. The fuselage was covered in ply and further strengthened by a sheath of glued and doped fabric. The wings were built from timber spars with fabric coverings, and hinged to fold at the centre. The Widgeon III was offered with single or dual controls, the seats being in tandem, with the pilot at the rear. The capacity of the standard fuel tank was 91 litres.

The Widgeon Mark III was slightly larger than the original design, having a wing-span of 11 metres and a length of 7.1 metres. Empty, it weighed 386 kilograms and laden, 635 kilograms. The manufacturers listed the top speed as 105 miles (169 kilometres) per hour, and the landing speed as 42 miles (67.5 kilometres) per hour.

Several were delivered to Australia. One was bought by Milton Kent, a promising young pilot and commercial photographer, who used it to win first prize at the Queensland Aerial Derby in 1927. Called the *Flying Cloud,* he purchased the Widgeon III mainly to use it for aerial photography, but following his success in Queensland, Milton Kent was offered the post as New South Wales agent for Westland. During 1928 he imported three Widgeons. One was bought by David Smith, who had previously

The *Kookaburra* showing the registration endorsement G-AUKA and a close-up of the script lettering of the plane's name.

owned the Ryan flown by Anderson. Smith died shortly afterwards in the crash of a Tiger Moth. The second was used by Milton Kent and still exists today. At the time of publication it is the oldest registered plane in Australia and is owned by Arthur Whittaker*, of Boort in Victoria.

When Keith Anderson received the unexpected gift from Smithy and Ulm, the Widgeon III had already earned a reputation as a lively and sporting little plane, which was able to take off and land in comparatively short spaces. He approached Milton Kent, who agreed to sell the third Widgeon, which was used as a demonstrator, for £889. Of the £1,000 gift from Smithy and Ulm, Andy had spent £400 on legal fees and other debts. Having no other money, he approached Smithy who signed a promissory note for £300, thus enabling the deal to be concluded on February 22nd, 1929. Anderson's receipt was written on the back of Kent's business card.

Andy named his new plane *Kookaburra*. It carried the registration endorsement G-AUKA. Unlike later models which had a split undercarriage, thus making them more suitable for rough field landings, G-AUKA had a straight axle undercarriage, with coil spring shock absorbers.

He planned to use the Widgeon to set a new world record for endurance flying in a light aircraft, and as soon as he took delivery, Andy sent the plane to a prominent Sydney engineer, Laurie Phipps, for the installation of additional fuel tanks, to boost the capacity to 430 litres. To carry the extra load, he asked Phipps to strengthen the undercarriage.

By April 1929, Anderson considered that he was ready for the record attempt. He applied for permission to use Richmond airfield which, he said, was the only take-off strip near Sydney that would be long enough for the fully laden *Kookaburra*.

Permission was granted, with the permit expiring on April 16th. For Keith Anderson, that day never came. Fate, which had treated him so miserably in the past, would deliver a fatal blow.

** The Westland Widgeon is now on permanent loan to Drage's Historical Aircraft Museum in Wodonga, Victoria. Owner of the museum, Mr Joe Drage kindly allowed Reg Morrison to photograph the plane for the cover of this book.*

The Kookaburra is lost

Undoubtedly the gift of money helped to rekindle the friendship between Smithy, Ulm and Anderson, but there was no attempt to re-establish the business partnership. Smithy and Ulm decided to 'go public' and start an interstate airline, which they called Australian National Airways. (This firm is not to be confused with a second airline of the same name, which is now part of Ansett Transport Industries.)

Neither Anderson nor Hitchcock was invited to join the new airline and this became obvious when Smithy and Ulm decided to fly to England in the *Southern Cross*. Ostensibly they were going there to order new planes for the airline, but they also wanted to complete their flight around the world, thus becoming the first to do so in the same plane. By this time, Harry Lyon and Jim Warner had returned to the USA. Despite this Anderson was not included in the crew once again. Smithy hired H.A. Litchfield as navigator and T.H. McWilliams as radio operator. As before, Ulm arranged the publicity and no less than 100,000 spectators descended on Richmond aerodrome to bid the aviators a tumultuous farewell.

Speaking from the cockpit, Smithy announced that the flight would prove that the British Empire could be linked by aerial communication. He passed the microphone to Ulm who added that they would bring closer the time when days, not weeks, would divide Australia from the hub of the world.

At 9.45 am March 30th, 1929, the *Southern Cross* took off carrying two and a half tonnes of fuel to fly non-stop to Wyndham in West Australia, on the first leg of what was to have been an historic journey. After 28 hours in the air, the *Southern Cross* sent out a short radio message:
'We are about to make a forced landing in bad country.' The nation listened for further word. None came. A day of complete radio silence passed and a wave of panic swept over Australia. Plans were made for a huge aerial search.

The sequence of events up to this point was fairly normal. Ninety minutes after take-off, the *Southern Cross* had lost its 90-metre long-wave aerial and could no longer receive messages, although it could transmit them. McWilliams had accidently unhitched a safety catch when opening a window to take a drift reading, thus causing the weighted aerial to unreel easily and

jerk free. When this happened, he sent a message to the cockpit:
'Long-wave aerial gone. Shall we return?' Smithy was never one to allow set-backs to discourage his plans. Besides, he had 5,000 litres of fuel on board and most would have to be dumped to ensure a safe landing. He decided to fly on, and replace the aerial at Wyndham.

Since he could not receive messages, Smithy was unaware that he was running into the grandfather of all storms. North of Alice Springs blinding dust, whipped up by the wind, turned into a sticky smear on the windscreen. With visibility close to zero, Smithy fought the plane as it lurched and bucked into the night. Even at dawn, Litchfield could not see through the gloom to take a sighting and, to make matters worse, Smithy had grave doubts about the reliability of the compass. The *Southern Cross* became hopelessly lost.

When the weather improved, Smithy found a mission station, (he later learned that it was called Drysdale). He dropped a note asking for directions to Wyndham and a group of natives, led by a white supervisor, lined up to form a pointer in the direction of south-west. Smithy waved acknowledgement, not realizing that the group below had failed to find his note. They were pointing to the nearest cleared land,

The flight path of the *Southern Cross* and the position of the forced landing.

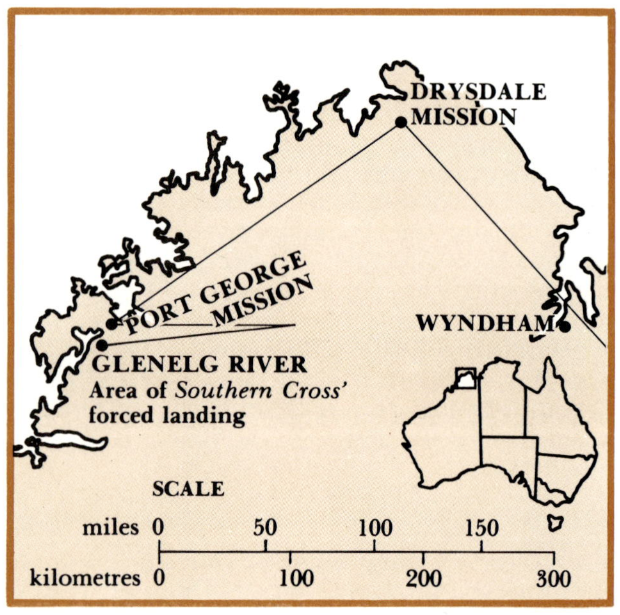

thinking that this might make a suitable landing strip.

He flew a considerable distance before ascertaining that he was not heading for Wyndham. After three hours, he sighted another cluster of buildings, later identified as Port George Mission. His fuel supplies were dangerously low, so he dropped another note asking the direction and distance to Wyndham. Large figures were laid out on the grass to tell him that there were 400 kilometres to go.

'We were dumbfounded,' Smithy said later. He headed east but after 20 minutes, he realized that he had insufficient fuel to reach Wyndham. He turned back, intending to land at Port George Mission, but when the fuel supply became critical, he decided to find a place to make a forced landing. A radio signal was sent: 'About to make a forced landing in bad country.' Eventually he found a level site — a mud-flat. Setting the plane on a landing course, he shut off the engines and glided down. The wheels of the big plane sank into the mud, but although it lurched and threatened to crash on

its nose, the *Southern Cross* finally stopped, undamaged.

After 28 hours in the air and one of the worst flights he had experienced, Ulm opened the locker. It should have been full of emergency rations — but he found that it was bare; the contents had apparently been removed at Richmond.

After searching the plane, the men assembled their total supplies — a few sandwiches, four kilograms of Allenbury's baby food, which was being carried to Wyndham by special request for the postmaster's youngster. There was also one kilogram of coffee, a packet of biscuits and a flask of brandy.

They brewed some weak coffee and laced it with brandy. Smithy joked about drinking Coffee Royal — and the name became permanent. The miserable forced landing and its tragic aftermath became known as the *Coffee Royal Affair*.

Although the aviators could pick up radio signals, they were effectively cut off from the outside world. Their transmitter was dependent on power from a small propeller which was normally driven by the wind as the plane flew through the air. Smithy and Ulm tried to arrange a belt-drive by jacking up a wheel and attaching a belt made from those which held up their trousers. This provided a little electrical

An aerial view of the *Southern Cross,* stranded at Glenelg River, Western Australia. Smithy's emergency landing in such difficult terrain was a brilliant achievement. The plane was virtually undamaged and no one was injured.

A close-up view of the *Southern Cross* shows the difficult terrain on which the forced landing was made. Smithy later flew the plane out, after fuel had been brought to the location.

power, so McWilliams devised an aerial and sent out some weak messages. They were never received. Ironically, Smithy and his crew could follow the search which was being made on their behalf, but they could not contact the searchers.

The disappearance of two national heroes created a wave of dismay across the nation. As the facts became apparent, it was widely assumed that the *Southern Cross* had crashed in alligator-infested swamps. If this were not the case, went the argument, the crew would surely have sent a second radio message.

As speculation grew from rumours, it was revealed that a signal had been sent to the *Southern Cross* as it approached central Australia, advising Smithy that he was heading into a severe electric storm. No acknowledgement had been received. After 24 hours of radio silence, the Civil Aviation Department asked West Australian Airways to commence a search. Using its terminus at Derby as a base, the airline dispatched a DH 50 immediately, and followed this with two more.

Public agitation mounted as no further news was heard of the missing aviators. Sydney's Commissioner, John Garlick (who had been appointed to temporarily head the city's administration), formed a Citizens Committee to raise funds and mount its own search. The response was so great that the sum of £4,000 was collected within hours. Soon the figure had risen to £7,000. The committee hired a six-seater

De Havilland 61 named *Canberra* and its pilot who was Les Holden. Holden was a fine aviator with an excellent war record, but even he was not able to move as quickly as the public demanded. The big plane with its 450-horsepower engine and laden fuel tanks proved too heavy to take off from the waterlogged strip at Mascot aerodrome. After draining some of the fuel, the *Canberra* was flown to Richmond, where the strip was longer. There was further evidence of official mishandling and public anger rose sharply.

The Prime Minister responded by ordering the seaplane carrier *HMAS Albatross* — then being refurbished at Sydney — to depart carrying six aircraft to help search the northern waters and rivers. There were inevitable delays as the crew was assembled and the ship refuelled. By the time *HMAS Albatross* steamed out of Sydney Harbour bound for the search area, 12 days had passed since the disappearance of the *Southern Cross*.

Whilst stranded on the river flats, Smithy and Ulm tried to improvise a method of driving the radio generator, using the aircraft's landing wheels. The resulting radio signal was too weak to be picked up. Had the signals been successful, the *Kookaburra* would probably have remained in Sydney.

Meanwhile, Keith Anderson was watching developments and informing his friends at Sydney's Customs House Hotel, that the planes were searching the wrong area. He was convinced that Smithy would have landed in the Port George area, because he had carefully followed the radio signals of the *Southern Cross*. After some days he persuaded the hotel's owner, John Cantor, to give support to a private search. Anderson said that he knew the area extremely well, having flown over it for some years with West Australian Airways. Further, he maintained that his plane was capable of staying in the air for 24 hours without refuelling and he also knew a reliable mechanic who was willing to go with him. Cantor agreed and together with some friends, he offered a sum of money to pay for the fuel and for Hitchcock's wages. In return he asked to receive any news of the search first, since it was probably his intention to try and sell the news to the Sydney newspapers.

Anderson's motives for initiating the search are unclear. Very likely, he wanted to repay Smithy's kindness or perhaps he saw himself as a public hero; the man who would rescue the famous Smithy from certain death. Perhaps he felt that finding the missing *Southern Cross* represented his last chance to vindicate his rather unsuccessful career in aviation and also create a new bond with Smithy. Success would also mean a new life for himself as a celebrity.

Irrespective of his thoughts, Anderson was not motivated by money.

'There is no foundation for the belief that Anderson would make any money out of the trip,' John Cantor said in a Press statement at the time. 'In fairness to him I do not feel justified in divulging the amount of guarantee that was subscribed by his friends. That amount however was a very small proportion of the sum which the Citizens Committee has in hand for the other three planes engaged in the search. Out of it, Anderson has to defray the cost of the insurance of the plane, payment to his mechanic, supplies of oil and petrol and any incidental expenses which might arise on the trip. I would say that when all his expenses had been defrayed, he would be very lucky to find himself out of debt ... Anderson had nothing to gain from the agreement. He and Hitchcock undertook the flight from a spirit of sympathy and fellowship for Kingsford Smith.'

It is certain that Anderson's preparations were hasty in the extreme and — possibly because of the financial situation — the plane was drastically short of emergency supplies. Apart from a flight from Mascot to Richmond, the plane had not been tested since the larger fuel tanks were installed for his proposed endurance flight.

The Customs House Hotel in the late 1920s. John Cantor was the owner from 1923 for 12 years and it was here that the necessary funds were raised to finance Anderson's flight to search for Kingsford Smith, the *Southern Cross* and the other crew members. The hotel which stands in Macquarie Place in Sydney, has been refurbished inside, but the facade remains almost the same.

However the most serious deficiency was the compass. For some reason which was never explained, the *Kookaburra* did not have a compass when it was purchased. Hitchcock owned a unit which had once been in the *Red Rose*, the plane flown to Australia by Captain W.E. Lancaster and Mrs Keith Miller. They had left England in October 1927, arriving in Australia the following March and the plane was subsequently sold to Mr J.R. Palmer. When finalizing the deal, Captain Lancaster had asked if he could keep the compass, as he owed his life to it. As it happened, Lancaster owed Bob Hitchcock some money (for services rendered), so he gave him the compass as a security for later payment. Hitchcock passed it to his solicitor, who may have put it in a steel safe, which would have affected its magnetic properties.

When Hitchcock agreed to travel with Andy on the *Kookaburra* he collected the compass and asked Laurie Phipps (who had fitted the larger fuel tanks), to install it. Later Mr Phipps said that the compass had been fitted at Mascot and 'swung' at Richmond and that Anderson was perfectly satisfied with it. He added that it was only one degree out.

It is probable that Andy knew the compass was faulty before he left Richmond, but this did not worry him. All his life he had navigated the routes by following telegraph lines, roads and railway lines and there was no reason to think that this journey would be different.

The absence of a suitable compass was not the only deficiency in the *Kookaburra*. There was no radio, little food, a meagre amount of water and a limited supply of tools. Bon was one of several people who felt that Andy should not make the trip. She was so adamant in her belief, that she used her most powerful argument. Bon told Andy quite firmly that if he went in search of Smithy, he could regard their engagement as terminated.

Nevertheless, on April 4th, 1929, the same day as Holden's *Canberra* eventually left Richmond, Anderson and Hitchcock climbed into the *Kookaburra*. The small group who waved them off included Bon Hilliard and two pilots — Milton Kent (who came in a Widgeon similar to the one he had sold to Andy) and a friend, Colin Ferguson. Kent later stated that the *Kookaburra* was very heavily laden with a gross weight of 1,975 pounds, (898 kilograms) which was 395 pounds (180 kilograms) overweight. Despite this, the *Kookaburra* 'had taken off spendidly', he said.

It reached Broken Hill that night and Andy sent Cantor a telegram reading:
'Arrived here safely 6.55 Sydney time, half hour after sunset. Experienced strong head winds all way, low misty clouds west of Blackheath necessitating landing Blayney for two hours. Been in air nine hours without refuelling, engine and machine functioning perfectly. Unless lucky enough to get following wind tomorrow will only be able reach Oodnadatta instead Alice Springs. Sorry — regards Keith Anderson.' (See page 55 for map.)

Anderson was telling only part of the truth. He had undoubtedly set off for Broken Hill but had unintentionally veered to the left near Bathurst and had landed to check where he was. Two employees of Blayney Motors happened to be driving down the main street. One was salesman Dick Jackson, the other an apprentice mechanic named Len Bell. When they saw the plane circle as though looking for somewhere to land, they followed. The *Kookaburra* touched down in a field, and they drove over to see if they could help. As a former war-time pilot, Dick Jackson was delighted to meet the aviators. He noted that the *Kookaburra* had run into a small

Hitchcock and Anderson with the *Kookaburra*, at Richmond, New South Wales prior to departure on their final journey.

rock, thus damaging the tail-skid. Len Bell was asked to straighten it out, while Dick Jackson drove the fliers to his home. There his wife prepared a hot meal and made up a packet of sandwiches to sustain them on the next leg of their flight. Noticing how cold Bob Hitchcock was, Jackson loaned him his old leather flying jacket which, according to Bell, still carried blood stains from the war.

The trip from Blayney to Broken Hill passed without incident, although Anderson was concerned about his erratic compass. After taking off the next morning and heading for Alice Springs, Anderson only managed to reach Marree, since he had run into enormous navigation problems. His telegram sent to Cantor from Marree, read:

'Departed Broken Hill nine twenty this morning arrived here after eight hours non-stop*. Experienced head winds at noon when sun in zenith and therefore unable obtain direction. Encountered large magnetic dust storm, compass going forty five degrees wrong which, with smallish map, caused big diversion from correct course. Was not able to establish position till sighting Port Augusta*, which has spoiled objective Oodnadatta today. Taking many practice bearings with inaccurate compass coming from Augusta and am easily able to steer correct course. Leaving here dawn, will spend tomorrow night Alice Springs. *Kookaburra* and engine functioning perfectly, petrol consumption four and a half gallons hour, oil one pint same** — Keith Anderson.'

Meanwhile regretting her hasty threat, the unhappy Bon sent a loving telegram to Alice Springs saying, in effect, that the engagement should continue. Fifty years later she learned the fate of her telegram. Dick Smith, whilst researching the *Kookaburra* story, discovered that the precious paper was listed among Andy's effects. He advised her accordingly.

Regardless of his fiancée's objections, Andy was determined to find Smithy. He set out for Alice Springs as planned and had only flown as far as Algebuckina, south of Oodnadatta, when the engine's running noise abruptly changed. Hitchcock was in the forward seat just an arm's length from the engine so he could see the exposed tappet on one cylinder vibrating badly, as a lock-nut came loose. He may have signalled

* The cross-country distance is roughly 400 kilometres.

* This meant that he had travelled due west instead of north-west.

** 20.46 litres of petrol per hour and 0.57 litres oil.

50

These photographs were taken by Norman
Bignell at Blayney soon after Anderson landed
there to ask for directions. The photograph of
Anderson, at right, is probably the last one
taken a few days before he and Hitchcock
became lost in the desert.
Anderson landed in a field, below, and the local
inhabitants quickly gathered to see the
Kookaburra and offer assistance, bottom.
The damaged tail-skid was straightened and the
aviators were treated to a hot meal and then
given a packet of sandwiches to take with them,
on the next leg of their journey.

Andy to descend, but that was probably unnecessary as the engine was running on 75 per cent power, which may not have been enough to keep the heavily laden plane airborne.

It was later reported that a prospector, Martin Kromer, had said that he was waiting with friends when an aircraft with a spluttering engine dropped down near him. He was amazed to find that the pilot had only a vague idea of where he was. He said that when Anderson asked the mechanic to get the tools, Hitchcock shook his head and replied, 'They've been stolen in Sydney'. Martin Kromer claimed that he saw Hitchcock adjust the loose tappet using a chisel for a screwdriver and the end of a corkscrew as a light hammer.

Irrespective of the tools used, the ingenious Hitchcock fixed the tappet and Andy took off, giving a 'cheery wave' as he did so. At Alice Springs he found a telegram from the Department of Civil Aviation telling him not to proceed. He ignored it, reasoning that he was on a private and not an official flight. Instead he sent Cantor this telegram:
'Arrived here safely at five twenty five from Marree, via Algebuckina where landed for hundred minutes make engine adjustments. Engine proving highly satisfactory. Think am out of head wind permanently. Secured large map tomorrow's route also much valuable information endeavouring make non-stop here Wyndham if Smithy not found by tonight. Feel fairly confident *Kookaburra* will pick him up Thursday, regards — Keith Anderson.'

The decision to go directly to Wyndham indicates that Anderson had a near obsession to reach Smithy first. He may have been racing Holden's *Canberra*, for the honour of finding the missing plane, or perhaps he had visions of Smithy and his crew dying of thirst. He could have been motivated by the opportunity to make headlines, but this will never be known. One thing is certain. When the *Kookaburra* left Alice Springs at 7.35 am on April 10th, it was grossly overladen with fuel but poorly stocked with provisions. There were four bottles of water, amounting to about three litres and the food consisted of two loaves, cut into mince sandwiches, and a small amount of cake. There were few tools; no axe, spade or similar equipment to use in an emergency landing. However there was an enormous quantity of fuel, as Andy had taken on 400 litres at Alice Springs. As a result, the *Kookaburra* was 195 kilograms overweight at take-off — and it had never previously been flown with such a load.

The prudent course was to follow the telegraph line and Anderson did so for the first 160 kilometres until he reached Woodford Crossing. Then he made an astounding decision. Despite the faulty compass and unreliable engine, he decided that he would reduce the time of the trip by a few hours if he diverted across country. Anderson was only carrying a general map of the area, but he knew of several important land marks. These included Ryans Well, Woodford Well and a rough shed on Lander River. From there he should pass over a salt-marsh north of Mount Davidson and, then take a straight line to Wyndham. Anderson had left word at Alice Springs that he might attempt to pick up the telegraph line at Halls Creek, as he was familiar with that part of the country. He also knew that if he flew south, he would find the defined road from Tanami to Halls Creek. However his immediate objective was to find a salt-pan known as Duck Pond, which was marked on the map, and keep it to the left. He flew on until he sighted a salt-pan, which he apparently mistook for Duck Pond. Confident that he was on course, Andy proceeded to curve in a northerly direction.

As Bob Hitchcock sat in the front seat, he saw the tappet slowly vibrating loose and since the plane was airborne, his horror would have mounted by the minute. He was in the agonizing position of knowing that engine power would rapidly fail as the tappet worked loose and he was in the perfect place to watch every movement of that tappet. As the power diminished, the heavily laden *Kookaburra* lost height. Anderson was forced to come down on an area of loose sand and thick turpentine scrub. He achieved a commendable landing in the circumstances, causing only minor damage to the undercarriage.

Anderson had no way of knowing, but he was a long way off course; their position was 128 kilometres south-east of Wave Hill station.

Bob Hitchcock soon had the engine repaired but this was the least of their problems. The tangled turpentine scrub, which had looked relatively innocuous from the air, was in fact, far too thick to permit a take-off. Working laboriously in the heat of the day, the men cleared a short runway but when they tried to use it, the sandy soil prevented the narrow Palmer Cord tyres from effecting enough speed for a take-off. Sharp remnants of burnt turpentine bush punctured the thin casings. In the gruelling sun, they set about extending the runway using their bare hands and a dual-bladed pen-knife. They then started a fire, perhaps in an attempt to attract attention or to burn off some of the scrub. The fire eventually burned out an area of 10 square miles (2,590 hectares).

The DH 61 *Canberra*, flown by Captain Les Holden who found the *Southern Cross* after it had been lost for 13 days. Holden also set out to find the *Kookaburra* but by that time Lester Brain (in the Qantas plane, *Atalanta*) had spotted it.

While Andy and Hitchcock were painfully extending the runway the eyes of the nation were still focussed on the plight of Smithy and his crew.

Despite a massive search, there was still no sight or sound of them. But on April 12th — two days after the *Kookaburra* had made its forced landing — Les Holden, in the *Canberra*, sighted the *Southern Cross* some 400 kilometres from Wyndham. At 9.50 am local time, the *Canberra's* radio operator, John Stannage (later to be closely identified with Smithy and his exploits) tapped out the message: 'Found. Found. All Safe.' His signal flashed across the nation like a joyous tidal wave.

Next day, the *Sydney Morning Herald* summed up the national feeling when it reported:

'Captain Holden and his observers in the *Canberra* yesterday morning located the *Southern Cross* on mud-flats on the banks of the Glenelg River, about 20 miles* south of the Port George Mission. All the crew are safe. The four men walked out in the open in response to signals from the *Canberra*. They appeared fit and well. Food was dropped from the *Canberra*. A running creek, near where the *Southern Cross* is located, provides an abundant supply of pure water. Expressions of congratulations at the finding of the crew of the *Southern Cross* have been received from all parts of Australia, as well as from Britain, different parts of the British Empire and the United States of America.'

The jubilation was short-lived. John Cantor broke the news that the *Kookaburra* was overdue on its flight to Wyndham. He appealed to the *Southern Cross* Citizens Committee to initiate a similar search. There was no mistake about his meaning when he added, 'The lives of Anderson and Hitchcock are as valuable to the state as any others'.

A very distressed Mr Cantor told the Press: 'Mr Anderson, before undertaking the flight, approached me and said that if someone found the money he would assist in the search for the missing plane and its crew. I agreed to finance him and Mr Anderson took upon himself the responsibility of the flight.

'I was invited to a meeting at the Town Hall this afternoon. I went to the meeting and, after giving certain information, I was informed that the Citizens Committee could not deal with the matter in any shape or form. I was asked how much I was prepared to guarantee to get a plane to search for Mr Anderson. I think my responsibility has ceased.'

Meanwhile one of the most extraordinary episodes in Australian aviation developed. Whilst the crew of the *Southern Cross* had eaten crawling shellfish, while waiting for the search party, Anderson and Hitchcock tore at the scrub, trying to extend the runway. A mass-circulation magazine, ironically called *Smith's Weekly*, claimed that the forced landing had been a publicity stunt. Rumours swept the nation, causing an uproar. Following, are the summarized allegations:

- Kingsford Smith had been handed a 'mysterious' note before leaving Richmond aerodrome.
- Ulm was reported to have said, on several

32 kilometres.

occasions, that there would be immense publicity if Smithy and he were lost in the desert. This would make it easier to get sponsorship for future record attempts.

- It was claimed that Kingsford Smith had been financing his old mate Anderson, who would become a national hero if he rescued the missing aviators.
- Anderson had waited until the search gained maximum publicity before joining in. He had told a friend in the Customs House Hotel exactly where he intended to look and that his predictions had proved to be remarkably accurate.
- No radio signals were received from the stricken *Southern Cross,* even though it had not been damaged in the forced landing.

A subsequent inquiry completely exonerated Smithy and Anderson from any such complicity. Evidence was given that the 'mysterious' message at Richmond was simply confirmation of a telegram which he had sent on the previous day, relating to weather conditions. Smithy explained the loss of the radio aerial adding that several attempts had been made to send signals from the mud-flat, but that as the plane was not in the air, they had no satisfactory means of driving the generators.

He also said that the crew had lit smoke signals but despite these, search planes had twice flown over them. Concerning the allegations about financing Anderson, Smithy angrily shouted that it was a vicious lie. He had given Anderson £1,000 after the court case and later endorsed a promissory note for a further £300, 'purely as a matter of friendship', he said.

In retrospect, it is patently obvious that if the forced landing was a hoax, it had been very badly planned. But the most convincing argument was that Smithy could not have known that violent storms would drive him off course. Had he arranged for Anderson to 'find' him on the banks of the Glenelg River, as alleged, it would have been a near miracle for him to fly through atrocious weather, become lost twice, operate with a faulty compass and finally descend at an appointed landing spot with fuel tanks almost empty.

●● ●● ●●

The one key witness who was not present at the inquiry was Keith Anderson. Later it was

After being rescued by Captain Holden in the *Canberra*, Ulm and Smithy were welcomed by Colonel Mansbridge and Mr Alex Scott, chairman of the West Kimberley Roads Board. Smithy had just learned that Anderson and Hitchcock were missing and was about to leave Derby to join the search.

discovered that he had briefly described the landing in the Tanami Desert on the fabric of the *Kookaburra's* rudder. In part, it read:
'10/4/29 to -/4/29. Force landed here 2.35 pm 10th April, 1929, thru push-rod loosening, No. 2 cylinder cutting out (as at Algebuckina, SA on 9/4/29 but temporarily fixed K.V.A.) exhaust valve and 25% h.p.

'Cleared bit of runway here which turned out just insufficient, or engine coincidentally lost power. Since 12/4/29 all efforts of course same next to nil, thru having no water to drink except solutions of urine (with oil, petrol, methylate [spirits] from compass). [Efforts] directed on obtaining sufficient power from engine to permit successful take-off.

'No take-off able to be attempted since 11/4/29 due increased debility from thirst, heat, flies and dust.

'Left Stuart (Alice Springs) 7.15 am local time and followed telegraph for 100 miles*, which was intention. Cut off then direct for point between Wave Hill and Ord River Downs. On a/c cross winds and inaccurate compass, and having practically only sun for guidance as large map showing only featureless desert...'

Anderson made five or six attempts to take off from the strip which they had painfully hacked out with a pen-knife and their bare hands, but it was too short. A tyre had been punctured, so they fitted a spare tube, which also became punctured. Since the cleared runway was too short and the plane had a flat tyre, more attempts at taking off were useless. Although no puncture repair kit was found on board later, it was assumed that Hitchcock would have known of the common practice employed by motorists, of stuffing a flat tyre with grass and leaves. Presumably, he was too weak to wrench the tyre off the rim.

Hitchcock was probably the more distressed of the two. He had been ill with blood poisoning, a result of the Bristol crash late in the previous year, and although the wound had been treated at Alice Springs, further attention was needed.

Whilst two incredibly lonely aviators struggled to survive, a minor battle was raging in Sydney over whose responsibility it was to look for them. The Citizens Committee was suspicious of John Cantor and, when he declined to advance money to conduct a search, they questioned him closely on the *Kookaburra's* route. On April 16th, they sent him a letter which read in part:
'Referring to your interview with the committee on Thursday last, at which interview you stated that Keith Anderson was to land at cattle stations en route to the north from Alice Springs

and to your subsequent statement to the Press that Keith Anderson was to fly direct to Wyndham and yet again your published statement that Keith Anderson would fly direct to the lost aviators at Port George, I am directed by the committee to ask if you are yet prepared to disclose to the committee the exact nature of the secret instructions you issued Keith Anderson?

'The committee is determined to do all in its power to find the missing airmen, but is hampered in its efforts by lack of information of the intended movements of the airmen, because what you termed your secret instructions, have not been disclosed,' signed J. Garlick.

Cantor shot back an angry answer:
'You either designedly or accidently misinterpreted my statements. I said Anderson may stop at a cattle station if he thought he could collect information on the way; his last wire from Alice Springs clearly indicated however that he was flying direct to Wyndham. I have never at any time stated that he would fly direct to the lost aviators, and I am disgusted to think that you should use such a statement. It is with pleasure that I note your committee is now prepared to do all in its power to find Anderson.'

Evidently disturbed by the altercation and the lack of action, Bon Hilliard's father sent a telegram to Colonel Mansbridge of the Air Board instructing him to 'despatch pilot O'Dea to search for Anderson in the north-east'. Colonel Mansbridge replied with a cable to the committee asking by whose authority Mr A.V. Hilliard acted. The honorary secretary replied that Hilliard was Bon's father but the committee had no knowledge of the suggested search by O'Dea.

Meanwhile, Smithy's airline, Australian National Airways, offered assistance and the committee accepted, saying its fund would defray the cost of fuel and lubricant.

More incessant worries followed complaints from Anderson's mother and Hitchcock's wife (both were living in Western Australia), that they had not received any news. It was decided that as an act of courtesy the Defence Department (located in Melbourne) should keep them informed by telegraph.

The next move came from the Prime Minister, who let it be known that 'he was much disturbed by a report that is circulating that the Citizens Committee in Sydney was disinclined to assist in the search for Anderson. They understood that the Commonwealth Government's support for the *Canberra* was to rescue Kingsford Smith and party and did not extend to the search for Anderson.'

The Prime Minister sent word to Mr Kitto of the committee saying that he wanted it 'clearly understood that if the *Canberra* can do anything to help in the search for Anderson, he was most

---

* *160 kilometres*

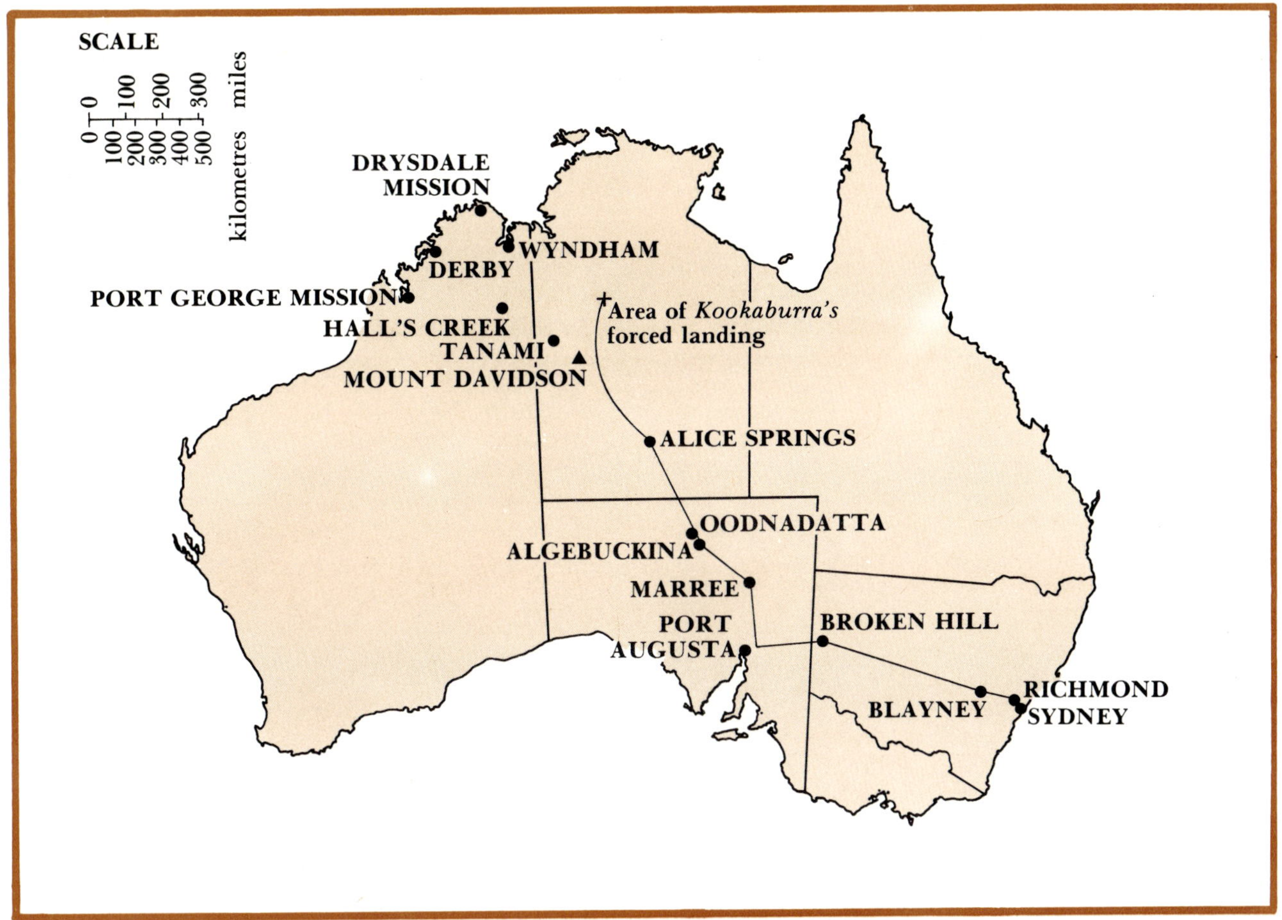

Map shows places mentioned along the route which Anderson and Hitchcock flew and the area where they were forced to land.

desirous that necessary action should be taken and that the Commonwealth Government would stand behind the committee in the matter'.

Within hours a search for the *Kookaburra* was under way. Five Royal Australian Air Force De Havilland DH 9A biplanes under the command of Flight Lieutenant Eaton left Laverton. Unfortunately the prolonged operation of these obsolete machines (they were designed as war-time bombers) took its toll. The 1918 Liberty engine in Eaton's DH 9A seized when they were close to Tennant Creek and although he was able to make a safe crash-landing, the plane was wrecked. The other planes all experienced problems, as the propellers tended to delaminate in the hot, dry conditions. By this time the Citizens Committee in Sydney had arranged for Les Holden to fly the *Canberra* towards Alice Springs and explore Anderson's route on the way. The authorities also appealed to the Queensland airline for help. A Qantas DH 50J, named *Atalanta,* was hastily fitted with a radio and dispatched with Lester Brain at the controls. It flew to Newcastle Waters, bound for Wave Hill.

Between Newcastle Waters and Wave Hill, Lester Brain spotted a column of smoke rising from the desert to the south-west. Diverting to investigate, Brain sighted an aircraft. He was soon able to read the lettering, G-AUKA. Realizing that he could not land anywhere near the site, Brain descended even lower and spotted a man prostrate under the starboard wing. He dropped some water then after radioing the location, Brain left to arrange a ground party to search for the other body. Lester Brain had located the *Kookaburra* at 12.10 pm local time, on April 21st, 1929.

The Kookaburra is found

Queensland and Northern Territory Aerial Services was a young airline known as Q.A.N.T.A.S. Although only seven years old by 1929, the airline was well organized, enthusiastically run and comparatively well equipped. The small, privately run organization had considerable expertise in long distance taxi-trips, which included flying over interminable stretches of barren desert. In later years the name was changed to Qantas Airways Ltd.

The pride of the early fleet was a De Havilland DH 50J, constructed under licence by Qantas and called *Atalanta*. According to the *Australian Motorist* of June 1st, 1929, it was 'equipped with a Bristol Jupiter engine of 450 horsepower. Possessing a cruising speed of nearly 110 miles per hour*, she is the fastest and safest long distance taxi-plane operating in Australia. Mr L.J. Brain, as pilot of the

* *176 kilometres per hour.*

Atalanta, is well known as having unique experience of flying conditions in the interior.'

Lester Joseph Brain was one of the most capable pilots of his time, with considerable experience of flying over deserts. Four years previously he had been chosen to fly an expedition across the inland to look for a supposed gold outcrop. Born in Sydney in 1903, he had joined the Air Force (later the RAAF), as a cadet, then in 1924 Brain applied to Qantas for a position as a mail pilot. After a short time he became chief pilot.

Qantas was, at that stage, operating a regular scheduled service on the route from Charleville to Longreach to Cloncurry, when news was received regarding the missing plane, *Southern Cross*. Within days, news arrived that the *Kookaburra* was also missing. Lester Brain was particularly concerned as he had met Anderson and Hitchcock during their around-Australia flight.

Above, Lester Brain and below, meeting Bob Hitchcock during the flight around Australia, which Hitchcock and Anderson made in 1927.

Left, the Qantas DH 50J *Atalanta*, photographed at Brisbane in 1929. Pilot L.J. Brain is standing in front of the plane.

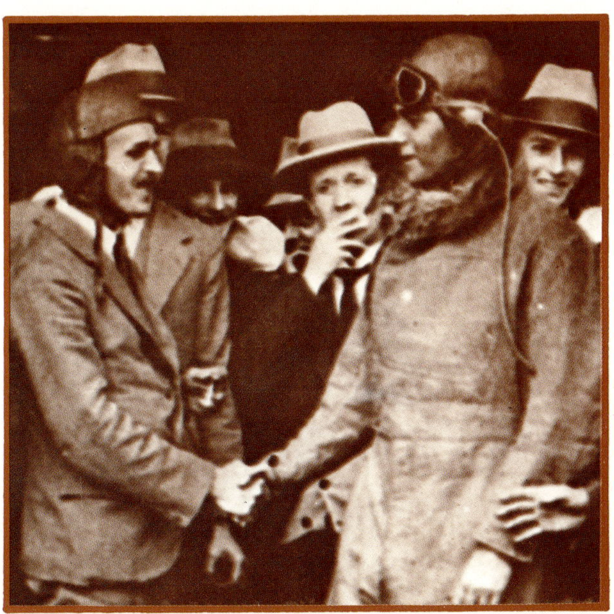

As the RAAF DH 9As departed to search for the *Kookaburra*, the Citizens Committee immediately called for more planes to help. The Defence Department and Air Board gave formal instructions to Qantas to join the search, for which they would be paid on a mileage basis. Hudson Fysh (later Sir Hudson, the celebrated Qantas chairman) who was operations manager at the time, quickly selected Lester Brain and the *Atalanta* for the task.

The speed with which the plane was prepared for the flight indicates the high degree of efficiency apparent in the young airline. Brain was in Brisbane at 5 pm on April 18th when he received his instructions. He immediately called Mr Robinson, a director of Queensland Radio Service, and said, 'Could we have a wireless in the plane?' Then he departed to supervise the fuelling of the *Atalanta* and make sure that all possible emergency supplies would be carried.

Like most aircraft of the time, the *Atalanta* was normally flown without a radio. Brain wanted one in case they were forced down in the desert and also to contact the base if and when they sighted the *Kookaburra*.

Mr Robinson telephoned F.W. Stevens, chief engineer of 4QG, in Brisbane, to say that Queensland Radio Service had been asked to install the equipment and would Stevens help. Stevens organized the necessary parts, then volunteered to operate the radio in the plane, taking his own receiver and a transmitter owned by his employer, Mr McIntosh. The receiver was evidently a powerful one, as he had used it a few days earlier to listen to the signals sent out by the *Southern Cross* on its ill-fated flight to Wyndham.

The transmitter was less powerful. Stevens described it as 'a very low powered affair, using nothing but an ordinary valve and a small bank of batteries'. He went on:
'The collection of spare parts entailed a lot of running about. I had to get certain radio shops to open up and let me have parts from stock...

'McIntosh and Alf Bauer, also a radio operator, went with me to Eagle Farm where the *Atalanta* was in a hangar. The Qantas people were busy on the plane and it was a very difficult matter to get near her. She had never had a wireless in before and a hole had to be made in the floor to take the aerial. A weight for the aerial was made out of a fitter's hammer head. It remained in position until we were over Roma when it was carried away.

'My previous experience in flying was almost negligible. I had had one or two short flights, but had never handled a wireless apparatus in the air before.'

Incredibly, Brain was able to set off at 6 am on April 19th, only 13 hours after receiving his initial instructions. In addition to Stevens, the

58

Atalanta carried P.H. Compston, an engineer.

'We hopped off a few minutes after 6 am,' said Stevens, 'and as soon as we got in the air, I let down the aerial and started transmitting.'

Apparently Brain did not experience any problems and, at 2.30 pm, the plane put down at the Qantas headquarters at Longreach for lunch and to take on the remaining provisions. Next day, Saturday April 20th, they left at first light and flew past Camooweal and on to Brunette Downs. After a hesitant start, the newly constructed wireless was working perfectly.

On Sunday April 21st, the *Atalanta* left Brunette Downs at 7.30 am and reached Newcastle Waters 80 minutes later. They were airborne again at 11.05 am, destined for Wave Hill where the local cattle station was a very sizeable property owned by a well-known firm of meat exporters. They had a reliable radio which was capable of transmitting and receiving. Later Wave Hill station became the centre of the RAAF operation.

By this time, the *Southern Cross* had been found and was ready to take off again. Smithy and his crew had been following the search closely for their former colleagues. In his book *The Old Bus*, Smithy said:

'We were completely worn out and in poor health and required medical attention but two days later we were sufficiently recovered to fly the *Southern Cross* to Wyndham, from which point we proposed to join the search.

'But just before we were able to leave Wyndham, the news came through that the *Kookaburra* had been found by Mr Brain in the *Atalanta* on the morning of April 23rd.'*

In fact the date was April 21st.

The Qantas pilot's narrative went on: 'We had gone about 50 miles* west from Newcastle Waters,' said Mr Brain, 'when I picked up a sign of smoke in the south-west, about 60 miles away. I knew this was desert country and there was no habitation and that blacks rarely visited here owing to the lack of water. When still 25 miles from the smoke — it was very flat country — I saw a dark brownish burnt patch and when within five or six miles of the edge, I saw an object on the ground. It did not seem to belong to the desert. It was something strange. It was the *Kookaburra* and appeared to be undamaged. We circled around at a height of 15 feet** and saw a body lying under one of the wings. I was satisfied that the man had been dead for a considerable time. We looked for the other man but there was no sign of him. On the remote chance that he might still be alive, we dropped a half a gallon† of water by parachute and also some food. I considered my best plan was to fly on to Wave Hill, from where a ground party might be sent across country.'

* *80 kilometres.*
** *4.6 metres.*
† *2.25 litres.*

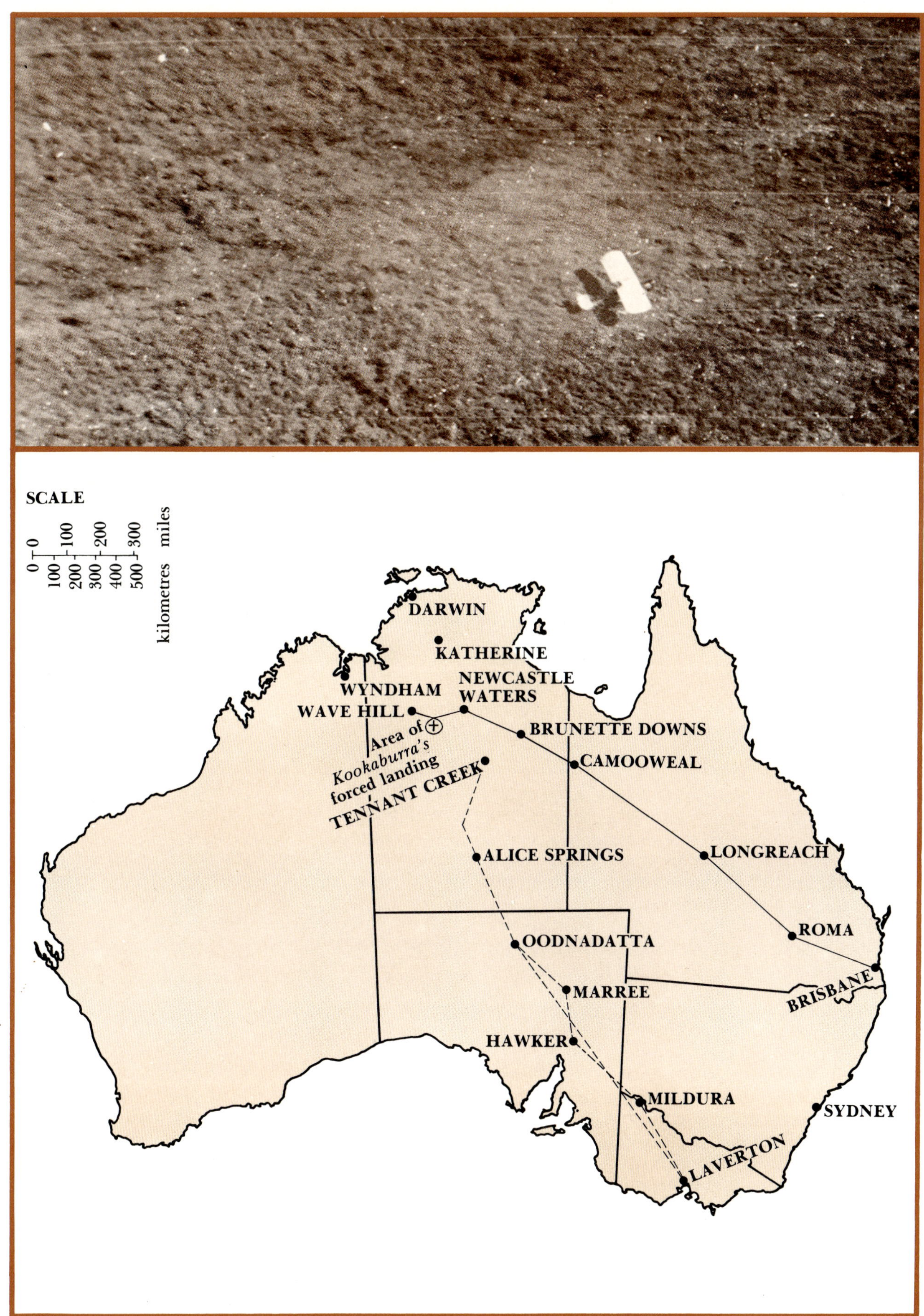
SCALE
0
0
100
100
200
200
300
300
400
500
miles
kilometres
DARWIN
KATHERINE
NEWCASTLE
WATERS
WYNDHAM
WAVE HILL
BRUNETTE DOWNS
Area of
Kookaburra's
forced landing
CAMOOWEAL
TENNANT CREEK
ALICE SPRINGS
LONGREACH
OODNADATTA
ROMA
MARREE
BRISBANE
HAWKER
MILDURA
SYDNEY
LAVERTON

The entire nation had been watching the progress of the search. One man, Captain Matheson, had set off from Goulburn in a Gipsy Moth to help find the *Kookaburra*. He only reached Cloncurry where he was firmly advised not to proceed because of the danger awaiting a solitary flier.

The RAAF had been pursuing a relentless search. Its planes — war-time DH 9A bombers — were led by Flight Lieutenant Charles Eaton, a tall, lean, sunburned young man with an easy smile, known to his colleagues as 'Moth' Eaton. Unlike the *Atalanta*, whose air-cooled engine was untroubled by the searing heat reflected from the sand, the DH 9As developed overheating troubles and could not fly continuously at low altitudes.

A fascinating account of the RAAF's role was outlined in a diary written by Flight Lieutenant Eaton and a separate one by Sergeant (later Group Captain) Eric Douglas.

'When the *Kookaburra* became overdue and no news was received of their safe landing,' Douglas wrote, 'an aerial search was ordered by the Minister of Defence, Sir William Glasgow. Within two days of their disappearance, two Royal Australian Air Force DH 9A two-seater biplanes, each powered by a 400-horsepower water-cooled Liberty engine, were on their way from RAAF, Laverton, Victoria...'

Eaton and his companion plane set off but were delayed when a leaking petrol tank in Eaton's DH 9A made it necessary to land at Oodnadatta. He taxied along the main street and parked near a pub. According to Eaton's unpublished account:

'It seemed that there was only one way to repair the tank and that was to remove it from the aeroplane. On account of our desperate hurry ... the whole township of Oodnadatta, not too many, just got to work with us in taking off the wings, the centre section, removing the tank, soldering a new seam to the tank, putting the tank back and repairing and re-erecting the aeroplane. We were away again 17 hours after landing at Oodnadatta.'

When Eaton wired the Air Board asking for extra help, three more DH 9As were despatched. On April 16th, he and Flying Officer Gerrard arrived at Alice Springs. In his diary, Sergeant Douglas, one of the pilots in the three planes which followed, continued:

'We put on board the usual cross-country kit of emergency rations, water bottles, hand tools, tying down gear, a few engine spares, some materials for air frame repairs and a first aid outfit. In addition, we stowed blankets, a service rifle and ammunition, a spare landing wheel and a spare propeller. The spare propeller was lashed securely to the fuselage ... where it could be readily seen in flight. It turned out later that I

Above, Flight Lieutenant Charles Eaton, who led the five RAAF DH 9As which searched for the missing *Kookaburra*. He was ordered to organize the ground party which buried the dead airmen.

Above right, Sergeant Douglas took this photograph of his DH 9A at Alice Springs on his way to the search. The identity of the men is not known.

Right, two of the men who found the lost *Kookaburra* — L.J. Brain (left) and P.H. Compston (right). In the centre is Mr Robinson, a director of 4QG, Brisbane, who organized the radio equipment for the search plane, *Atalanta*. Robinson's deputy, F.W. Stevens, accompanied Brain on the flight.

was carrying the only spare wheel and spare propeller... great difficulty was found in obtaining maps from Mildura onwards in our proposed route and in fact they did not exist in any detail over most of our journey and search area.'

Pilot Officer 'Man' Allen was in charge of one plane accompanied by leading airman 'Snow' Allen. The third was flown by Flying Officer Ryan accompanied by Doug Endean, a wireless operator.

62

Only one plane was fitted with 'wireless telegraphy'.

Sergeant Douglas provides some marvellous glimpses of the navigation difficulties involved: 'No strip or aeronautical maps were available for use in our cockpits,' he wrote, 'and we made sketches for use on our hull pads showing the relevant points and distances, this being worked up from several large maps of Australia which we carried.'

The planes flew at about 1,200 metres, partly for visibility reasons. 'It also gave a good altitude in case of engine trouble and made it easy to control the temperatures of our engines in the cooler air,' said Douglas.

'We had no elaborate instruments to check on wind speeds and its direction. To enable us to maintain a good course we reverted in the main to the simple method of checking on our drift by back bearings and land horizon observations.'

The three planes had no difficulty in finding Hawker and they set course for Marree which would take them more or less to a railway line, thus providing easy navigation. At one point, the pilots suddenly found themselves blotted out by thick dust extending from ground level up to 1,800 metres. They battled through and finally arrived at Marree with scarcely any fuel. All pilots were forced to switch to their small reserve tanks and a check on one showed it contained only nine litres.

The pilots personally refuelled from large drums, using special strainers to exclude moisture. After a quick conference, the pilots decided to fly across Lake Eyre, straight to Oodnadatta, rather than follow the longer route taken by the railway. Although they would be flying over unmapped areas, the men considered the risk worthwhile, as the reduced distance gave them a chance of reaching their destination before dark. At first they tried flying low, but the engines began to overheat, thus forcing the pilots to climb to cooler air. As the evening with its cooler temperatures approached, they were able to descend to lower altitudes again. The reason for this, said Sergeant Douglas, 'was to get our eyes accustomed to the depleting light in readiness for a landing on a strange aerodrome'.

Flying Officer Ryan had the fastest plane and was first to sight the aerodrome, which he indicated by rocking his aircraft. He went ahead, landed and whilst his comrades circled above the field, Ryan put down a strip of white cloth to show the approximate touch-down point. By the time the others were ready, the light had failed, so a fire was lit. The pilots came in safely, only to find the strip was very rough and almost surrounded by tall camel-pens. Ryan's plane had struck one of the pens on its approach, thus damaging a tail-skid, but this was repaired.

They learned that Eaton and Gerrard had landed safely at Oodnadatta, after repairing the leaking fuel tank, and had set off for Alice Springs. Next morning, Douglas and his colleagues refuelled. After driving camel-teams and a few Aboriginal spectators off the landing strip, the three planes took off for Alice Springs, where they easily found the small landing field. Flight Lieutenant Eaton gave them a hearty greeting and told them how the search was progressing. Soon Flying Officer Gerrard touched down, having completed his reconnaissance. Eaton briefed the newcomers and said that, where possible, the planes were to conduct the search in pairs.

Sergeant Douglas set about servicing the engine. As he laconically reported in his diary: 'Unfortunately, the engine work was rendered difficult due to the loss of our workshop kit of tools, which had been thrown overboard near Oodnadatta from our leading aircraft. When asked to give a reason, the pilot said: "Don't bother with those things, when your engine stops, you stop"!'

Due to the plane's limited fuel range, Eaton sought an open area of land near Ryans Well and used it as an advance base to increase the search radius. Petrol was carried to the base from Alice Springs by truck. Two aircraft were sent to Tennant Creek and, when Eaton telegraphed the pilots, they said that on the way they had sighted a bad fire but could not investigate because neither plane was totally serviceable.

The search started in earnest on April 20th. The pilots had prepared message bags which they dropped at homesteads and settlements asking if a mono-plane had been seen and, if so, when and in what direction was it flying? Instructions were given on the method of replying. They were to lay a white cloth on the ground. Douglas thus learned that 10 days previously, the *Kookaburra* had passed over Morton's Homestead, north of Ryans Well, flying in a north-westerly direction.

The next day, Douglas was flying in company with the planes flown by Eaton and Gerrard, when he noticed white smoke coming from Eaton's engine. Flying closer, he could see molten aluminium dropping from the engine; a sign that a piston had seized. He says that he quickly dived down and picked out the most suitable spot to which Eaton could glide. He ascended to the ailing plane and signalled the direction of the suggested crash-landing site.

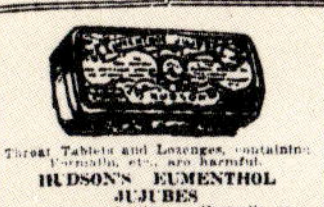

The Daily Guardian

NET SALES, 169,613
"Smith's Weekly" Net Sales Exceed 206,000 Each Issue.

FOUNDED BY JOYNTON SMITH

TO-DAY'S WEATHER: Mostly fine. **1D**
(Special City Forecast.)

Vol. VI., No. 252 — Registered at the G.P.O., Sydney, for transmission by post as a newspaper. — MONDAY, APRIL 22, 1929 — Telephones: Editorial and Business, 8 lines—B7111 to B7118; Publishing, B5691, B5692.

GRIM ENDING TO SEARCH FOR KEITH ANDERSON

KOOKABURRA DISCOVERED BY Q.A.N.T.A.S. PLANE

"LYING IN PATCH OF BURNT GROUND ... APPARENTLY DEAD BODY, BELIEVED TO BE ANDERSON, BENEATH WING ... MACHINE UNHURT ..."

BUT NOTHING SEEN OF HITCHCOCK

LOCATION; NORTHERN TERRITORY, BETWEEN POWELL'S CREEK AND WAVE HILL, IN DREADFUL COUNTRY

GRASS AND SCRUB BLAZED TO BRING RESCUE

Grim tragedy marked the end of the search for Keith Anderson and his plane the Kookaburra, yesterday afternoon.

The Qantas plane Atlanta, flying low, found the Kookaburra in dreadful country north of Wave Hill station in the Northern Territory.

Anderson is apparently dead; his body lying at this moment beneath the wing of his plane, after twelve days of indescribable horror.

Of Hitchcock there is no news—did he march alone into the desert to try and reach civilisation, while Anderson stayed by the plane, lighting smoke signals?

The Air Board has already instructed five Air Force machines, now at Tennant's Creek, to proceed to Newcastle Waters first thing in the morning to search for Hitchcock.

A land party is setting out from Wave Hill to the Kookaburra.

BRAINS' OWN STORY

Captain Brain was guided to the Kookaburra by a fire which had cleared an area ten miles square, started by the ill-fated airmen. The man seen lying under the plane's wing had evidently been dead some days, as he was black in the face.

There were indications that the Kookaburra had to land owing to engine trouble. The cowl of the engine was removed, and attempts made to clear a run-way for the machine to take off again.

Anderson was found 200 miles north of the route he was supposed to take. His compass error would account for this.

FIRST official news of the discovery came in a message from Wave Hill, where pilot Brain landed after locating the Kookaburra.

The Qantas aeroplane located Anderson's machine in the desert, approximately 80 miles west by north from Powell's Creek.

"The aeroplane circled low, and Brain saw a man lying under the wing, apparently dead for some days. He looked like Anderson.

"The aeroplane cowling was off. A party with a tracker is setting out from Wave Hill without delay. Brain anxious that other machines should join him to make a close search for the missing man."

BRAIN'S OWN STORY

(The following message is direct from the Brain):

WAVE HILL, Sunday.—"Anderson's machine is in the worst part of the desert, about half-way between Powell's Creek and Wave Hill wireless station. It is apparently undamaged.

"Engine trouble, it seems, has been the cause of a forced landing.

"On the ground near the monoplane the area for ten square miles and is still burning.

"It will take several days for a camel expedition to reach the spot.

"This will have to be directed and kept supplied by aeroplane, owing to the barren nature of the country which will have to be traversed.

"It seems clear that Anderson and Hitchcock endeavoured to make a run-way for their machine.

"To-morrow, at daylight, I leave for Newcastle Waters to pick up the five R.A.A.F. machines. Together we will search the country beyond Anderson's machine and endeavour to locate the other man. I consider there is no chance of finding Hitchcock alive."

HITCHCOCK NOT SEEN

("GUARDIAN" SERVICE.)

WAVE HILL, Sunday Night.—The Atlanta left Newcastle Waters at 11.30 a.m. Sighted smoke N.W., 12.15 p.m. Flew over the plane, which was apparently undamaged. A man was lying face down under the wing. It looks like Anderson. The plane dropped water and tinned food. There was no sign of the other man.

A short runway had been cleared. The engine cowling was on the ground.

The Atlanta arrived at Wave Hill at about 2 p.m. eastern time.

Pilot Brain searched for an hour, looking for Hitchcock, but without result.

From appearances, Anderson had been dead some days. He was black in the face.

Apparently a smoke fire had been set up by the aviators, and started a bush fire, which cleared an area ten miles square.

It will take two or three days to organise a proper ground expedition.

It should be possible for the ground party to quickly clear a landing ground near the machine, provided water and food are kept up to them by dropping from aeroplanes.

BROKEN BY GRIEF

MR. CANTOR HEARS OF TRAGEDY

Mr. John Cantor, who was Lieutenant Anderson's backer in the flight, was a broken man last night when the news reached him.

He has lost a friend in whom he had the utmost faith. Up to the very last he believed that Anderson was safe.

Now, with his hope shattered, he—

Bob Hitchcock.

FIANCEE BRAVE UP TO THE LAST

Brave and hopeful up to the last minute was Miss "Bon" Hilliard, fiancee of Lieut. Keith Anderson.

Only a few minutes before the tragic news came through she made the following statement to "The Guardian":—

"I am hopeful of hearing good news soon, now that the planes are working on Keith's immediate route after what has seemed a long delay in ascertaining particulars of that correct route.

"I have every confidence in his skill and resourcefulness, and feel sure that they safely crossed the desert, and that they are in an area far more north to what the search planes are doing such wonderful work."

IS IT KEITH LYING DEAD?

ONE FOR RESCUE, AND ONE WITH PLANE

PERHAPS A SMASH

(By CAPT. E. W. PERCIVAL.)

MESSAGES from the Qantas plane, piloted by L. J. Brain, leave many gaps which only examination of the ground and the plane will enable us to fill up.

The machine, we are told, appears to be undamaged, and is standing at the east end of a large patch of red ground, largely burnt and still smoking.

It is apparent at once that the pilot of the Qantas plane did not consider the landing ground safe. He had to fly low to make his observations and then made his best speed to Wave Hill station, 70 miles away, to send out a party on horseback to explore the spot and see what could be done.

It may be that when the military plane reach the spot to-day one of them may risk landing. The question of getting into the air will not be as important as it would have been had the Qantas plane taken the risk and been unable to take off again.

It will take two hours from Wave Hill station at least a day to reach the spot. If they do six miles an hour it will be fast going in that country. It may take them longer.

The reasons for the man lying under the wing may be one of several. He may have been thrown out and killed, and his companion may be dead within the machine, or directly underneath.

It is more probable, however, that both men were uninjured when the plane landed. Then they would decide that one of them should go away and endeavour to reach a station to bring help, or meet with natives, while the other stood by the plane to light signals. In the end, when exhaustion and the want of water took toll, the one standing by the machine would naturally take shelter under the wing from the sun, and would there become helpless there.

If one of them went for help he may have been the waterless desert from any objective, and too far from his companion to return when he became exhausted.

BURNING OFF

The fact that the ground around the apparently undamaged plane is burning seems to point to the fact that the man whose body lies beneath kept fires going as long as possible, in the hope of attracting attention. That the grass is still smouldering may mean that the man is not yet dead, but is too weak to move, because it cannot be so many days since the grass was lit, and apparently kept under control.

The phrase in the message, "it looks like Anderson," takes added weight in regard to identity, because Keith would probably stand by the plane. He was the pilot, and would be better able to assist in getting the machine away if help should come from an unexpected quarter. In that case he would have to take off and assist in the search for his companion.

It is impossible to say yet what happened, but it looks very like a forced landing as a result of some mechanical trouble.

Maybe this trouble was overcome and the men were trying to clear a space for taking off by burning the grass, but found that the task became too great for them and they gave it up.

The one grim sentence in the message is that which says that the man has "apparently been dead for some days." To those who know the North-West this may mean much more than the observer cares to write.

HOPE FADING

Later messages report that a runway had been cleared, and that the cowling of the engine was on the ground. This indicates a forced landing from mechanical trouble. It would seem, however, that attempts at repair were either futile or exhaustion overcame the aviators before the repairs could be completed. Apparently the mechanic (Hitchcock) worked at the engine while Anderson attempted to clear a space in which to take off.

Want of food and water defeated these heroic efforts, and the final story is told by the pathetic figure lying under the wing. Alone, and with blackened face, the silent form tells us of the grim toll thirst and hunger exact in the waste and inhospitable spaces of the great North-West.

CROSS RECALLED

HOLDEN WILL AWAIT INSTRUCTIONS

MELBOURNE, Sunday.—Defence Department has wired the Southern Cross to return to Sydney via Gordon Downs, Tanami, and Morton, and to drop messages recalling ground search parties.

Captain Holden, of the Canberra, has been advised to return to Wave Hill and await instructions.

HUGE TRACT OF DESERT

KOOKABURRA'S FINDER KNOWS COUNTRY

LOOKED FOR GOLD

WHEN Mr. J. L. Brain, Brisbane manager for Qantas, set out on the search he said he was determined to find Lieut. Keith Anderson.

And he did, but under tragic circumstances.

So far as is known, only two white men have penetrated the huge tract of desert country in which Anderson and Hitchcock crashed.

One of them is Mr. Brain, who in 1925, with a companion, searched for gold in the immense desolation.

Pilot Brain

"The country is covered with turpentine scrub and giant spinifex. There would be no choice of country for Anderson to land in," said Mr. Brain before his departure on the search.

WATERLESS WASTE

"If forced down Anderson would be lucky not to have his machine damaged. Further, he would be lucky if he found food and water. There are only isolated water holes in this scarcely-known country.

"One travelled for over 100 miles without a sign of water," said Mr. Brain.

"From a flying point of view it is very bad country in which to meet with a mishap. For 300 miles after leaving Alice Springs there is not a landing place. The first one to be met with then is about 30 miles south-west of Wave Hill."

An R.A.A.F. plane was forced down yesterday, eight miles from Tennant's Creek, owing to engine trouble.

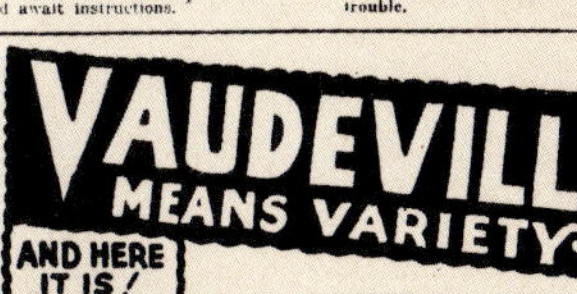

MISSING SINCE APRIL 10, Keith Anderson's plane Kookaburra, is now reported found down in the north, with the intrepid pilot who went in search of the Southern Cross believed dead.

There was no sign of the other man, R. S. Hitchcock.

"Brain dropped a can of water attached to a small parachute.

"Anderson's machine appears to be undamaged, and it stands at the east end of a large patch of red ground which has been largely burned off, and was still smoking.

WHERE KOOKABURRA LIES

Map of the country where the Kookaburra was found.

The Daily Guardian

NET SALES, 169,613
"Smith's Weekly" Net Sales Exceed 200,000 Each Issue.

(FOUNDED BY JOHNSON SMITH)

PUB. DAYS. WEATHER: Mostly fine.
(Special City Forecast.)

1D

Vol. VI, No. 255 — Registered at the G.P.O., Sydney, for transmission by post as a newspaper.

THURSDAY, APRIL 25, 1929

INFAMOUS ATTACK BY "SUN" ON ANDERSON

VILE ATTEMPT TO BLAME DEAD

SWITCHING BRUCE'S SEARCHLIGHT OFF THE SOUTHERN CROSS FIASCO

10,000 WRITS FOR £10,000 CANNOT CHOKE PUBLIC OPINION

CALLOUS LEADING ARTICLE

DURING the past few days, the Sydney "Sun" and its "Telegraph" satellite were in obvious fear of public contempt. But the lowest depths had to be reached. The leading article in the "Sun" yesterday was infamous.

News published in "The Guardian" yesterday morning made it clear that Mr. Bruce was bent upon investigating the recent flight fiasco with its tragic sequel, but had not yet decided the scope of his inquiry.

Immediately the "Sun" made a desperate effort to belittle and blacken the heroic adventure of Keith Anderson, and to whitewash the errors in the voyage of the Southern Cross.

Everybody knows that the "Sun" and "Telegraph" were financially interested in Smith and Ulm's flight.

The "Sun's" shameless attempt to provoke condemnation against Anderson (who is dead) and to arouse approbation for Smith and Ulm (who are living) was a vile debasement of a newspaper's opportunities.

The mean and caddish endeavor to escape the due inquiry was beneath all public decency.

Not one word in the "Sun's" leading article yesterday suggested that inquiry should be made into the flight of the Southern Cross.

But the "Sun" was indecently eager to cast stones at the ghost of the Kookaburra. Let us quote some of the "Sun" sophistry:—

"These two brave and unfortunate young men"—Anderson and Hitchcock—"went out on their quest of their own free will."

—"Sun" leading article yesterday.

NO WORD OF THIS

But no word to say that they went out, for no other reason save that the Southern Cross was sensationally lost, and mysteriously silent. Not a hint from the "Sun" that Anderson would never have started, but for the limelight of "public anxiety" round the names of Smith and Ulm.

It was an independent attempt, suddenly conceived and hastily equipped ... less carefully organised than the flight of the Southern Cross.—"Sun" leading article.

At least Keith Anderson carried a set of tools, which (as the "Sun" well knew) was more than the vaunted organisation of the Southern Cross could boast.

The suggestion that an inquiry be held into this ill-fated flight is one with which everybody must concur.—"Sun" leading article.

Everybody has concurred and is still concurring, and Mr. Bruce has decided, that inquiry must be held into two flights. Not only the Kookaburra flight, in which Anderson and Hitchcock died, but also the Southern Cross flight, which induced them to go to their death.

Nobody save the "Sun," with its special financial interest in the matter, has tried to prevent the inquiry from covering the flight of the Southern Cross.

MOST CRUEL OF ALL

In the "Sun's" leading article yesterday there is one sentence which will probably remain unrivalled for its callous cruelty:—

"If they" (Smith and Ulm) "forgot a few emergency precautions ... it did not involve them in tragedy."—"Sun" leading article.

Here any man must marvel, how a financially interested egotism can blind the eyes of even a newspaper corporation like the "Sun." It is true that no tragedy befel Smith and Ulm because of the forgetfulness their backers suggest.

But an awful tragedy did befall Anderson and Hitchcock, who hastened to the rescue; and, just because of that consequent tragedy, the public have craved the general investigation which the "Sun" sought in vain to suppress.

Smith and Ulm themselves cannot be blamed for the abominal callous attitude of the "Sun."

Their own personal grief for Anderson's death is surely too great to allow them, as fellow-flyers, to sneer at the dead in this superior manner.

SINCE the Southern Cross was discovered on April 12, at the Glenelg river flat, "The Guardian," on behalf of the Australian public, has asked for investigation [...] reasons why the big plane and crew of four remained lost for so long.

Simultaneously with "The Guardian," the Sydney newspaper, the "Labor Daily" boldly and loudly formulated a questionnaire, which contained solid facts for public inquiry, and the answering published that very day. We supported [...]

"Labor Daily" made a firm demand. One morning the "Sydney Morning Herald" supported the need for inquiry, but receded when Professor Payne (Air Accidents Investigation Committee) declined to initiate investigation.

"The reply to 'The Guardian' came from the offices of Minister ... and faith and of C.F.P. Ulm is the chaos of true writs of £1000 each.

"The Guardian" is not conscious of having published any libel. We welcome investigation.

GREETED HIM

Smiling Keith Anderson shaking hands with pilot Lester Brain when he and Hitchcock were welcomed at Brisbane Aerodrome when they arrived there on route around Australia on a commercial flight. Tragedy of which neither could have had any inkling has brought them together again.

BRUCE WILL BRING BODIES HOME

ANDERSON'S MOTHER IS GRATEFUL

WHAT MEMORIAL?

Keith Anderson and Bob Hitchcock will not be left lying in the lonely wilderness.

It was ordered yesterday by the Prime Minister (Mr. Bruce) that the Commonwealth Government would itself pay the last and honor, of bringing their bodies home for burial.

MR. BRUCE courteously telegraphed "The Guardian" yesterday:—

"Instructions have been issued to the Central Australian authorities to bring the bodies of Anderson and Hitchcock in to the nearest settlement, if this is possible.

"The bodies will then be sent to wherever the relatives desire them to be buried.

"As the Commonwealth will bear all costs, private subscriptions are now unnecessary."

McCORMACK PROMPT

Mr. W. McCormack, Premier of Queensland, was equally prompt and courteous in his answer to the appeal from the citizens of Sydney. He wired yesterday to "The Guardian":—

"I would willingly give every help possible to have the bodies recovered duly transported to New South Wales, to be buried wherever the relatives desire.

"McCormack, Premier."

It becomes unnecessary for members of the public of N.S. Wales to contribute further to the purpose formed yesterday by Mr. Percy Freeman and others—namely, to bring the bodies of the brave dead knight. Mr. Bruce has undertaken this mournful duty to the name of the nation.

Many subscriptions were received yesterday, with pathetic tributes which revealed the depth of public feeling.

SUBSCRIPTIONS HELD

No further subscriptions for the transport of the two dead flyers are now desired.

Subscriptions already received are being held by "The Guardian."

It may be that funds will be raised either for the comfort of relatives or for a public memorial.

In that case, if subscribers wish to authorise the transfer of their gifts to such a purpose, it can be done.

JUST A LETTER

Botany.
April 14, '29.

Good old "Guardian,"—

Just read paper after a sleepless night thinking of our two boys about to be buried away out back in the bush, a disgrace we would never live down.

Start to us and them, out insist on bringing them back where the mothers of Sydney can go and put a white flower on the spot where they lie.

I am not putting my money in the fund for that because I haven't a pennie in the world, but Friday is pay-day, and this is one cottage where cost of living will be cut down, and you will receive but a little for the memory of our wonderful brave dead.

We depend on you to see that you do what these two boys have done, and what we all must do.

"OUR BIT."

MRS. ANDERSON'S THANKS

Keith Anderson's mother to living in Perth.

Yesterday she was seen by "The Guardian's" representative in that city and she sent the following message to the citizens of Sydney:—

"I gratefully appreciate the splendid sympathy of all and the kindness to bring Keith's body to Perth."

From the Mayor of Goondiwindi (Q., Mr. K. A. Ferguson, came the following unexpected telegram to the editor of "The Guardian":—

"We not allow Anderson and son Hitchcock to be buried in the desert. It would be a public disgrace. I will donate on guinea towards cost."

BON HILLIARD'S COURAGE

"MY KEITH IS SO MAGNIFICENT"

WIRE TO MOTHER

Anzac Day to-day, and the saddest day in the life of Bon Hilliard, Keith Anderson's fiancee.

YESTERDAY she gave up hope at last. But she gave it up bravely. She smiled with a broken heart. The smile for Keith's gallantry.

"He is so magnificent," she said, and her gaze looked far away through tears of memory.

They had been such companions in the days gone by. Last to say goodbye to Keith and Bob Hitchcock when they left for England was Bon, with her ever-cheery smile. Last to farewell them on their fatal search was Bon smiling.

And in her sorrow she still smiles—"My Keith is so magnificent." For "WAR"—is—and ever will be.

DID NOT FLINCH

When her father, Mr. A. V. Hilliard, had told her that hope could no longer be held out, she did not flinch. She turned to her mother. "We must tell Keith's mother how we feel" for this telegram was sent to the broken, lonely mother in West Australia.

Do not grieve. We must be brave. Keith is magnificent.

Not "war"—is, and ever will be. We are being overwhelmed with sympathy," said Mr. Hilliard yesterday. "All day long the phone rings and rings. [...] people come to the door of our home of ...vereance We are so deeply honored, so inexpressibly grateful.

"You know, it seems so bare and cruel that Keith—in her best man any call—should have gone to his death.

"He is so magnificent."

MRS. ANDERSON ILL

PERTH, Wednesday.—Mrs. Anderson is still confined to her room under medical care, and only immediate relatives are allowed access.

WHERE PLANES ARE:

Atlanta (Qantas)	Longreach
Canberra	Commercial
Southern Cross	Alice Springs

ONE SEARCHER RETURNS

DARWIN, Wednesday.—Information has been received by Police Inspector Stretton to the effect that Kingsford Smith has reported having seen Mounted Constable Sheridan, who was searching for the Kookaburra, and having informed him that the plane had been found. Sheridan returned to Wave Hill.

AIR FORCE TASK

Following inquiries conducted by the R.A.A.F. in Sydney yesterday, this body will be entrusted with the task of getting the Kookaburra off the ground and bringing it on to Sydney.

RED CROSS WREATHS

Girls of the Junior Red Cross will to-day place wreaths at the Gates of Memory as a tribute to Keith Anderson. One in the shape of an aeroplane has been presented by Miss Jaques, Sorlof, Bondi Junction, and the other was prepared by the girls themselves.

CITIZENS' RESCUE FUND

Previously acknowledged	7,541
Lodge Lane Cove No. 232	1 1
Western Suburbs Returned Sailors and Soldiers' Club	2 2
Prof. Sir Edgeworth David	1 1
	£7,539

GOVT. ORDERS INQUIRY

INTO SOUTHERN CROSS AND KOOKABURRA

MANY QUESTIONS

("GUARDIAN" SERVICE.)

CANBERRA, Wednesday.
Prime Minister Bruce said to-day that the Government had decided to hold an inquiry into the forced landing of the Southern Cross, the loss of the Kookaburra, and also into the crashing of one of the Defence Department planes.

FOR the moment, the inquiry will be limited to this, as the Government had not yet decided exactly what form it would take

It would, however be given full power to obtain all the information it needed.

Although the Prime Minister did not announce it, it may be taken for granted that the inquiry will not be conducted by the Air Accidents Board, for the reason that three of the members of that board are members of the Government Service, and there has been considerable criticism of the department in connection with the search.

It is probable, too, that the inquiry will be much farther than the matters mentioned above.

The events of the last few weeks have emphasised the necessity for regulations dealing with flights over uninhabitable country in Australia, and to other countries, to ensure that before a plane starts on such a venture it will be properly equipped to meet all emergencies, and is in a proper condition to attempt the voyage on which it is to embark.

REGULATIONS

There are stringent regulations for ships in regard to seaworthiness and loading, and also to the apparatus for the saving of life in the event of error.

There are certain regulations with regard to aeroplanes, but the powers the Commonwealth possesses are apparently inadequate.

It is probable that when terms of reference are issued to the body conducting this inquiry, it will be found that they will be required to make a report on the existing regulations and recommendations for their improvement.

Mr. Bruce was asked about a suggestion that the Government make provision for the dependents of the two airmen, but replied that when information as to their circumstances was available, and there was an application for assistance, the matter would receive sympathetic consideration.

WEIRD ORGY OF EMOTION AT TOWN HALL

AGED WOMAN FALLS IN FRENZIED FIT

DIVINE HEALING!

Every bit as grotesque and revolting as his fervour of demonstration, was Victor Cromer's mass-healing meeting at the Town Hall last night.

HE retired to the back of the hall this time to switch on the so-called "Divine Power," but the effect was there, and practically the identical power-seekers.

Of course, there was a collection from a fairly well filled hall.

Rather more grotesque were the movements allegedly actuated by this power. A woman rode her chair like a racehorse, digging in the spurs and using the whip. Another pirouetted and twisted.

A man scratched himself ferociously while a red-haired girl made motions as though casting out a fishing line.

WOMAN HAS FIT

The climax was reached when an elderly woman, who had been swinging her shut fists like a prize-fighter, collapsed with a crash on the stage, in a fit. Immediately Cromer's woman assistant seized the opportunity to explain her symptoms.

The organ droned spasmodically throughout the seance. Suddenly an angular girl prostrated herself and "howled" continuously.

Cromer again urged the people to "tune in with God." Other equally incongruous phrases and Americanisms fell from his lips.

Other members of the audience performed to fret the power, and, convulsed erratic movements to the consternation of their neighbours.

A pathetic figure on the stage was that of a thin little child, who fluttered her hands trying her best to follow what was going on around her.

Next to her, on a mattress, writhed a woman who writhed, postured, and as included in an extraordinary manner. Her actions were remarkable.

If possible, last night's exhibition was even more revolting than Cromer's first demonstration.

SON ACQUITTED OF PARRICIDE

WOMAN FOUND GUILTY IN ERICA MURDER

MELBOURNE, Wednesday. — That Albert Wilson, 72, made a tragic error by taking the blame for the murder of his father, and showed great chivalry, was the plea Mr. Brennan K.C. submitted to the jury at the Criminal Court to-day.

Wilson and Anna Elizabeth Hilliard, 34, were charged with the murder of David Wilson, 53, at Erica, on March 12.

The jury found Hilliard guilty of manslaughter with a strong recommendation to mercy. Wilson was found not guilty and discharged.

Mr. Brennan, for Hilliard, made a strong appeal for leniency, and informed the Judge that she was in a certain condition. She was remanded for sentence.

CRUSHED BETWEEN TRAIN AND POST

RAILWAY CLEANER SERIOUSLY INJURED AT EVELEIGH

Caught between the end of a shunting carriage and some stone pillars in Eveleigh workshops last night, John Martin, 53, a cleaner, residing at 33 George Street, Redfern, was badly crushed and taken to the Royal Prince Alfred Hospital, where he was admitted by Dr. Jones with a fractured collar-bone, internal injuries and shock.

BRUCE'S ACTION

Mr. Bruce, before announcing the decision of his Ministry, has evidently weighed the whole matter carefully.

He is understood to have declared that the investigation must be conducted by Absolutely independent persons.

Herein the Prime Minister shows his shrewd insight.

Two persons who should not be placed on the investigating Board of Commission are Colonel Brinsmead (Director of Civil Aviation) and Mr. M. L. Shepherd (Secretary for Defence).

We make not the slightest imputation against the personal integrity of either of these two gentlemen.

But, in a recent public inquiry, an association was revealed between these and the managing editor of the "Sun," Mr. H. Campbell Jones; and the social rivalry which was the dominant at that inquiry makes it a matter of street propriety that Colonel Brinsmead and Mr. Shepherd should be stood aside.

"AGE" TRIBUTE

The Melbourne "Age," in its leading article yesterday, wrote ...

"It may be that the full story of this tragedy of the desert will never be known, but enough is already known for the people of Australia to realise that the body of a very gallant and noble-hearted man lies under the wings of the Kookaburra far out in the waterless desert. He gave his life in the attempt to save others. No man can display a nobler spirit of self-sacrifice than that. Australia mourns his death, but feels a spirit of pride in the fact that his lonely death in the trackless desert was the culmination of a splendid and heroic enterprise. ... To the list of heroes who have sacrificed their lives in the attempt to save others must be added the name of Lieutenant Keith Anderson, and in all probability that of Mr. R. B. Hitchcock, the mechanic who accompanied him in his flight."

FROM THE PEOPLE

This note will be placed on the Cenotaph to-day by Capt. W. G. Stroud, 20th Batt, A.I.F. and former company commander of his late "Bob" Hitchcock, sometime of the Kookaburra.

THE 30th BATTALION (W.A.) honor their glorious dead and add their old-time comrade. H. S. Hitchcock (Kookaburra) to the Battalion's roll of fame. Humbly,

W. G. STROUD, Capt. 30th Batt, A.I.F.

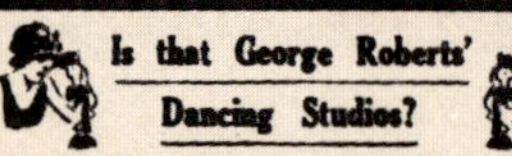

Mr. Bruce

This is the Westland Widgeon mono-plane
which was flown by Milton Kent, the New
South Wales agent for the plane in the late
1920s. Milton Kent arrived at Richmond
aerodrome in this plane to farewell Keith
Anderson and Bob Hitchcock when they took
off in search of Charles Kingsford Smith. This
plane is of the same design as Keith Anderson's
Kookaburra, having a wing-span of 11 metres, a
height of 3.2 metres and a length of 7.1 metres.
The inset shows the rear cockpit where the pilot
would sit.

The joy stick or control lever is in the middle
and the instrument panel shows four dials —
the one at right indicates revolutions per
minute, at the bottom is the oil pressure guage,
at left is the air speed indicator and centre top is
the altimeter.

This cockpit is the same as the one in which
Anderson would have been seated on his final
journey in the *Kookaburra.*

This Westland Widgeon is believed to be the
oldest registered aeroplane in Australia and is
owned by Arthur Whittaker of Boort in
Victoria. It is currently on display at Drage's
Historical Aircraft Museum in Wodonga,
Victoria where owner Joe Drage made the plane
available for photography.

David Poynter

David Poynter

The insets on this page show the making of Hitchcock Highway at top left. Top is the huge Caterpillar grader which was used to clear the tangled turpentine scrub.
Left, Dick Smith and Claypan the grader driver.

The large photograph shows the blow-hole or sink-hole which Pip and Dick Smith sighted from the air. This became a key point in finding the remains of *Kookaburra*. Lester Brain, who found the plane in 1929, told Dick Smith that he had seen a blow-hole on his way back to Wave Hill. In addition Dick had the distances travelled by the truck party after they had made camp for the night at the blow-hole on their way to the *Kookaburra*, so Dick was able to calculate the approximate position of the plane thus leading to its eventual discovery.

The inset shows Tony Peter looking into the centre of the blow-hole or sink-hole. It was caused by a collapse of the earth's crust due to a weak surface. A limestone cave system opens up below the surface, and this one was found to contain mud.

The remains of the *Kookaburra* found by Dick Smith's expedition in 1978. It was ravaged by bush-fires and the harsh weather. The insets show close-ups of the remains.

Above, the lower left-hand side of the cirrus engine of the Westland Widgeon III with the cowl removed. The heavy tubes are the engine mounts.

Below, the port wheel with its tyre is stuck upwards on its axle and struts. The strut with the coil spring is normally the rear strut so the view is from underneath looking forward. The large brown shapes are the extra fuel tanks which were fitted for the endurance attempt which Anderson had intended to make.

The light grey skeleton-like structure lying across the tanks is the remains of the hinged movable flap of the wing on the left-hand side of the plane.

Frank Scott

Eaton's DH 9A was wrecked when he made a crash-landing after his engine had severely overheated. Eaton and his co-pilot were not hurt. This photograph of the remains laying near Tennant Creek was taken some time later.

Later, Eaton recalled the incident in his diary, saying:
'I had no alternative but to land in small rolling hills along the telegraph wire. It was a good crash, the trees stripping my wings, pulled me up and I just flopped to the ground. The pilot of the other aircraft, seeing that I was not injured, went on and landed at the Tennant Creek telegraph line station. The mechanic, who was with me, and I started to walk into Tennant Creek, where we were met by a buckboard with two well-known central Australian identities, Messrs Woodroff and Martin, the linesmen of the Tennant Creek telegraph station.'

Eaton and his mechanic, Sullivan, were picked up and taken to join their colleagues at Tennant Creek where they heard that a Qantas plane, the *Atalanta,* was on its way to join the search. Flying Officer Ryan was despatched to Newcastle Waters to meet it. No sooner had he left, when a radio message arrived to inform them that Brain had sighted the missing plane on his way to Wave Hill.

The Air Board received the news which was tapped out from the *Atalanta* to Wave Hill

On page 64, the Prime Minister of the time, Mr Bruce, ordered that the bodies of Anderson and Hitchcock should be brought home for burial. The front page of the *Daily Guardian* in Sydney also gave support to Anderson after their rival newspaper, the *Sun,* had condemned his flight as being 'suddenly conceived, and hastily equipped'.

station. Brain advised that only one man — believed to be Anderson — was visible from the air, and he was laying under the wing. Instructions were issued for fuel to be sent to Wave Hill so the planes could continue looking for the other man. A total of 16,900 litres plus 1,730 litres of oil was loaded aboard a train at Darwin, consigned to Katherine, from where the fuel would be trucked to Wave Hill. Another message to Eaton said:
'Desire Air Force machines and *Canberra* and *Atalanta* to co-operate tomorrow in complete air search 40 miles* radius Anderson's plane. Also facilitate all possible arrangements assistance of ground parties.'

The Air Board had previously instructed the *Canberra* airliner (as it was called in those days), to take Dr Hamilton of Sydney to the nearest landing point to the *Kookaburra,* adding that 'all available aeroplanes will be massed today for an intensive search for the missing man'. Meanwhile Smithy, who was still in Wyndham, announced that he would also join the search for the missing aviator and would call at various stations on the way to inform them that the *Kookaburra* had been found. He also proposed to try and land near the plane to look for the missing man.

On receipt of his new instructions, Eaton sent a telegraph message asking Lester Brain to meet him at Newcastle Waters. He also decided to leave Allen's plane at Tennant Creek as the engine was irreparable.

Hence only three of the original five DH 9As set off for Newcastle Waters where they met Brain and the local authorities. It was decided that the DH 9As and the *Atalanta* should proceed to Wave Hill, via the site where the *Kookaburra* lay, which they should find easily

* *64 kilometres.*

because of the smoke from the bush-fire. Flying Officer Ryan reported that he had seen the smoke again on his flight from Tennant Creek to Newcastle Waters. He said that he had returned to Newcastle Waters with the intention of making a search in company with Mr Brain, and so 'lost the honour of sighting the *Kookaburra* first'.

The four planes soon came across the *Kookaburra* standing in blackened scrub. The pilots signalled each other that a landing anywhere near the plane would be dangerous. Sergeant Douglas later recorded in his diary: 'It was obvious that Lieutenant Anderson had executed a marvellous landing under most difficult conditions.'

The RAAF men had come prepared to parachute down, if necessary, with Eaton insisting that he would be first to jump if they found a man alive.

The four planes swooped low, flying around the *Kookaburra,* but the DH 9As were severely handicapped, as their water-cooled engines began to overheat. The *Atalanta's* air-cooled unit seemed impervious to the reflected heat.

The men could clearly see the dead body of one aviator under the wing. They searched the surrounding area but could see no other clues. As the DH 9As could not continue flying low over the desert, the four planes set course for Wave Hill, which they estimated to be about 144 kilometres to the north-west. About 29 kilometres from the *Kookaburra,* Lester Brain noticed a peculiar hole in the ground, resembling a natural blow-hole. It was between 12 and 15 metres deep, but no water was visible. Brain was delighted with the discovery and, on April 23rd, he sent a telegram to the Air Board, suggesting that the cone-shaped hole could contain water which would be of great help to a search party. Further on, the fliers noticed a large clay-pan which would have made a safe landing area, if only Anderson had been able to stay in the air long enough to locate it.

The group reached Wave Hill where they were met by the station manager and his wife. They also met Mr Moray, the pastoral superintendent for Vesteys Ltd, a major meat exporter, and owner of Wave Hill station. He volunteered to join the ground party proposed by Eaton, and make his car available for the trip.

Meanwhile the news that the *Kookaburra* had been found reverberated around Australia. The Sydney newspaper, *Daily Guardian,* contacted Qantas who directed Lester Brain to fly to Sydney, where he was interviewed.

Smithy also flew to Sydney. On the way he passed over the *Kookaburra,* which was still undisturbed, since Eaton's ground party had yet to arrive.

'We saw the machine lying there exactly as described by Brain,' Smithy later wrote, 'surrounded by a fire which was blazing on a five-mile* front. As we drew near, Ulm and I took turns at circling low down over the *Kookaburra,* within 15 feet** of her and we surveyed the awful ground on which Anderson had been forced to land. He had selected the only possible landing place for miles around and it would have been suicide for any other machine to have attempted a landing there. Even had a landing been made, it would have been impossible to take off again.

'We caught sight of a body lying under the wing of the plane, clad only in singlet and underpants ... The sight of the *Kookaburra* and the dead body lying there affected us deeply.'

Charles Kingsford Smith then flew into Sydney where, only a few weeks earlier, 100,000 cheering friends had given the *Southern Cross* a tumultuous farewell. His return was greeted with extreme hostility. Staggered by the abuse and insults hurled at himself and Ulm, Smithy tried to explain that the rumours suggesting that the *Southern Cross* had deliberately become lost were not true. The longer he talked, the louder became the shouting.

Public opinion condemned Smithy as a villian who had caused the death of two courageous mates. Although this view was widespread, the man who had found the missing plane disagreed.

Fifty years later, Lester Brain wrote to Dick Smith and expressed his astonishment at the extraordinary manner in which two men with a poorly equipped plane, few tools and a faulty compass had set out to cross the Tanami Desert after they had already made a forced landing due to engine trouble the previous day.

* *Eight kilometres.*
** *4.6 metres.*

Smith's Weekly **was quick to wring the drama and emotion from the tragedy, implying that there was more to the story than the public had been told. The question which the newspaper raised in everyone's mind was, 'Why did Keith Anderson go out to die?'**

CHÂTEAU TANUNDA HOSPITAL BRANDY

PRICE 4d

Smith's Weekly

The Public Guardian

FOUNDED BY JOYNTON SMITH

Net Sales Per Issue Exceed 306,901.
Sydney "Guardian" audited net sales 169,615 daily.

Vol. XI, No. 11 (Copyright) — Saturday, April 27, 1929

Registered at the General Post Office, Sydney for Transmission by Post as a Newspaper

CHÂTEAU TANUNDA HOSPITAL BRANDY

Questions To Be Answered

THE SACRIFICE!

Why Was It Necessary For Anderson to Have Crashed as He Did?

SOUTHERN CROSS FIASCO

DEAD ... dead ... dead ... that lonely body under the plane wing, with the smouldering scrub around him as a majestic funeral pyre ....

Dead ... dead ..., a heroic and immortal figure, sacrificed nobly to a noble and unselfish quest.

There is no man in Australia who will not bare his head. Any voice that speaks of Keith Anderson must quiver. No eyes can be certain that they will not shed tears.

Tragedy in the adventurous records of Australia has never been more terribly cruel than this. Why did Keith Anderson go out to die? What necessity called him?

All Australia with a grieving heart asks Why; and all the world looks for an answer to the question.

SMITH'S presents to its readers all over Australia the question which is asked by that mute figure in the desert solitude, with its scorched lips and sun-blackened face. WHY? Keith Anderson and Bob Hitchcock would never have set out on their fatal flight had not the disappearance and the silence of the "Southern Cross" been prolonged beyond all limits of time within which anxiety could be reasonably allayed.

Keith Anderson knew how skilled in flight, and how resourceful in emergencies were his old-time mates, Kingsford Smith and Charles Ulm.

When the forced landing of the "Southern Cross" was first announced, Anderson was among those who confidently trusted that Smith and Ulm would speedily be again in touch with the world ....

"Got To Go"

Keith was a "chooms" friend in Smith's office, and after March 31 he called more than once a day to ask anxiously whether the expected good news of the "Southern Cross" had yet arrived.

But, when day succeeded day, and still no word came from the missing crew, Anderson's anxiety increased. He feared that there had been some uncrossed calamity.

A friend was needed. Rescue was called for. He himself must be on the wing.

"By —!" he cried, with a profane comrade's oath which came from his heart, "we were mates together, and I've GOT to go after them!"

He was practically "broke"—so hard up that he had to borrow helmet, goggles, and tools.

If the accidents of Fate had not kept the "Southern Cross" and her crew so mysteriously silent, neither Anderson in his quest of Death, nor any other more fortunate rescuer, would have needed to start at all.

But the "Southern Cross" in what had looked like a simple stage of a well-ordered world flight, was crossed by misfortune in many ways.

The ironical tricks played by circumstance upon Kingsford Smith and Ulm constituted a "Southern Cross" Word Puzzle.

Why did all these troubles accumulate upon the "Southern Cross" in a supposedly easy flight across the Australian continent?

Why should this well-advertised flight have grown into a fiasco, which has already ended in this tragedy of Wildest Australia?

Aust- tralia — a tragedy that should never have been necessary.

Smith and Ulm confidently flew round Australia a little more than two years ago. More recently they flew across Australia in the "Southern Cross" without mishap.

What savage irony of ill-luck piled up deficiency upon deficiency against them in this last flight? Look at the links which broke in their expected "chain of safety."

Wireless Went Wrong. From Wyndham, on the day when they started (March 31) there came a telegram intimating that they must not start. Weather was bad. The Wyndham correspondent said in gave no terms:—

"Conditions unfavorable and unsafe, will be unsuitable for probably some days."

Nevertheless, the "Southern Cross" had started, and "could not be spoken to by wireless" because of some mishap to the receiving aerial. Smith and Ulm could not be advised to make an intermediate landing for safety's sake.

Nor, after they had landed on the Glenelg River, were they equipped to send out a message telling their whereabouts.

Hence the acute public anxiety, and hence the brave leap of Keith Anderson into the arms of Death. Why, oh why this unhappy turn of the wheel of misfortune?

Maps went wrong. When the "Southern Cross" did land, he was within twenty miles' bee-line from the Mission Station at Port George; and within twenty miles of walking distance which the men of Smith Head, covered in a day and a half!

But Smith and Ulm carried no map on which the Port George Mission was marked—"and so did not know in which direction to walk," says Kingsford Smith.

For want of a map on a map the crew was lost. For the loss of the crew Keith Anderson was lost. And for the heroic but dreadful fate of Anderson the—

"Smith's" will no longer ask of Fate the reason why. It will put the facts to those by whom the flight of the "Southern Cross" was organised.

For the sake of that sun-blackened body in the northern wilderness, all Australia will make inquiry and will demand the answers.

The "Southern Cross" set out, and she disappeared with her crew of four, thus equipped or unequipped —

She failed to receive a weather O.K. from Wyndham before starting.

She failed to receive an imperative recall after starting.

She lacked necessary equipment to repair her wireless.

She had no means of sending out a reassuring signal after landing.

She was without tools: not even a hacksaw, a hammer, or a tomahawk. "One screw-driver," says Kingsford Smith. "Two screwdrivers," says Ulm in his Diary.

She was without food— "seven sandwiches about to putrefy," says Ulm's Diary and later he produces an unnamed number of tins of gruel, which were being carried to a sick child at Wyndham."

Finally, whatever smoke signals the stranded party on the Glenelg River sent up, they failed to attract the notice of planes which searched their immediate vicinity for many days.

The Port George Mission had seen them fly past, and had given them signals and directions. But none of the distress "smokes" sent up by the "Southern Cross" crew could bring curious natives from in and around the Mission.

Ulm and Smith in their own narratives say that they saw their whole party threatened with early death from hunger and thirst, their only food being "snails and grass"—apparently whelks and a native bean—besides the gruel. And they lived through it.

Anderson and Hitchcock in their

(CONTINUED ON PAGE 9.)

"Just a Shape under a wing of a plane in Wildest Australia!"

"SMITH'S" VAUDEVILLIANS TAKE PART IN A SHOOTING AFFAIR

Eaton's ground party

Lester Brain and Charles Eaton had decided it would be too dangerous to try to land a plane near the *Kookaburra*, so Eaton sought approval to parachute himself nearby, in the hope that one airman may have survived. The Air Board advised against it, unless evidence could be seen from the air, that the missing aviator was still alive.

The Air Board instructed Eaton to organize a ground party using Wave Hill station as a base, to recover the bodies if possible or bury them on the site. Eaton told his companions that if the *Kookaburra* was undamaged, he would try to fly it out.

The ground party consisted of Eaton, Sergeant Douglas and Mr Moray, the pastoral supervisor of Wave Hill station. Three Aboriginal stockmen were chosen and there were 26 horses to carry supplies.

Moray offered to make his 1927 Buick tourer available, thus allowing the white men to go on ahead, whilst the stockmen followed with the horses. Air support was arranged, with Flying Officers Ryan and Gerrard and Sergeant Campbell detailed to fly the three DH 9As as overhead guides. A system of signals was devised to communicate their progress and the ground party's requirements.

Douglas was included in Eaton's group because he was a talented mechanic and could be needed to keep the car mobile and, later, to inspect the *Kookaburra*, if it proved possible to fly it back to base. Because of his experience as a bushman, Mr Moray was asked to lead the ground party to and from the *Kookaburra*.

Whilst these plans were being made, the Air Board sought permission from John Cantor and Bon Hilliard's father to move the plane. As the DH 9As were to drop food and other supplies to the ground party, Ryan was given the leading role in the air. His was the only DH 9A with a wireless and he was instructed to report all progress to the Air Board.

One of the first messages he received was a direction that Eaton should bring the bodies back to Wave Hill if possible. Word was also sent from Mr Hilliard giving permission for the *Kookaburra* to be salvaged and brought to Sydney.

Moray's car had a single bench-seat, so the three men squeezed in, as best they could. A strong wire guard was fitted to protect the radiator and the car was heavily laden with fuel, oil, tools, tyres and supplies; Moray had personally supervised the loading. Lester Brain stayed until the necessary organization had been completed then he farewelled his colleagues and, with his crew, flew to Longreach.

The expedition set out on Wednesday April 24th. For the first hour they made good progress along a rough track leading to a water-hole

An RAAF DH 9A.

Three of the Aboriginal stockmen, with some of the horses, prepared for the overland trek to reclaim Anderson's and Hitchcock's body.

called Junjiminji, where they had agreed to meet with the stockmen and horses. Travelling together, the party managed a further 16 kilometres before camping for the night.

By then it was obvious that the loose boulders and rough ground were damaging the Buick. The terrain covered on the second day, was worse. In his record of the event, Eaton described the country as a 'low brush, with a type of grass called turpentine, the tips of the brush being extremely brittle'. According to Douglas, Mr Moray was rather aggressive in his driving methods:

'Mr Moray set the car like a tank at the scrub and we moved forward in second gear and ploughed a pathway through it.'

Not surprisingly, the Buick was soon halted due to a hole in the radiator and two flat tyres.

While Douglas repaired the tourer, Flying Officer Ryan and Gerrard appeared overhead and signalled the direction of the nearest available water. Then they landed a short distance away on a clay-pan. Eaton and a stockman named Sambo went on horseback to meet them.

Ryan was asked to bring back a soldering kit and more food. Eventually, the car was mobile again but within a few kilometres, the radiator began to leak and the replacement tyres were flat. Eaton decided to abandon the car and proceed on the horses.

The Aboriginal stockmen located water, so the group established a campsite for the night and left the car in the care of one of the Aboriginals. The other two stockmen, Daylight and Sambo, then guided the party, displaying

their extraordinary sense of direction. In his diary, Douglas described how the Aboriginals knew each horse by its hoof marks and could track any horse, even when travelling on hard ground where there were virtually no impressions.

Douglas also commented on how surprisingly cold the nights were in comparison to the extreme heat during the day. The flies worried the men even more than the heat, but they struggled on. Soon one of the aircraft appeared, and the pilot signalled that the blow-hole — their main land mark — was just three kilometres away. They reached it, pressed on for a further eight kilometres at which point they were just 21 kilometres from their goal.

'It was decided,' recorded Douglas, 'to make a dash for the *Kookaburra* on the next day, as we were within striking distance, even though our horses would be at least 40 hours without water before they reached the billabong where we had last watered them. Mr Moray explained that the horses could go for 50 hours between watering, provided they were trained progressively, but this could not be achieved in the first long deprivation of water.'

The horses were hobbled at night but several strayed to seek water. The men rounded them up at first light, thus delaying their departure. The party forged ahead as quickly as the conditions permitted and, by mid-day they estimated that they were in the vicinity of the *Kookaburra*.

They soon discovered how difficult it was to locate the plane in the thick scrub. The burned area was more extensive than they had expected and Daylight and Sambo, although they were experienced trackers, returned without any idea of where the *Kookaburra* lay. The horses by this time were extremely thirsty and although hobbled, they were trying to escape to find

The plan for Eaton's ground party to drive to the stranded *Kookaburra* was abandoned when Mr Moray's Buick gave repeated mechanical trouble. This photograph was probably taken by Sergeant Douglas.

water. By mid-afternoon the search was temporarily abandoned. Mr Moray gave instructions that it was time to leave and they had almost reached the blow-hole when Ryan appeared overhead, almost on cue, and directed them to the blow-hole.

Although this incident was relatively minor, the repercussions appeared in headlines around the country. Ryan did not realize that the party had already been searching in the area where the *Kookaburra* lay, so he circled above them, trying to find out why they were not pressing on towards the missing plane as intended. A puzzled Ryan sent word to the Air Board who issued a statement saying that it was mystified as to why the ground party had made camp at the blow-hole.

The reaction of one Sydney newspaper was typical of most when it reported:
'The movements of Flight Lieutenant Eaton's land party which arrived at Anderson's plane yesterday are becoming mysterious. When Flying Officer Ryan flew from Wave Hill to the *Kookaburra* yesterday, he found the party 10½ miles* from the scene of the tragedy apparently on the way back to Wave Hill.

'Ryan did not clear up the position in his message to the secretary of the Air Board (Major Coleman) last night. His message is so obscure that it suggests the party may have run short of water and will return to the *Kookaburra* today.'

The party spent 36 hours at the blow-hole which Douglas described as a natural occurrence, about 7.5 metres in diameter, with fairly steep sides and about 15 metres deep. The reason for the delay was that there was no water in the blow-hole, so Sambo and Daylight were instructed to take the horses back to the billabong, 29 kilometres away and to return in the cool of the following evening.

Moray burned off a patch of tall grass, to make their camp position more discernible to the DH 9As. During the night, Douglas arose to relieve himself and, in the dark, fell into the side of the blow-hole and severely wrenched his

* *16.8 kilometres.*

The *Daily Guardian* published this front page when the ground party was only 10 miles (16 kilometres) north-west of the *Kookaburra*. It showed an artist's impression of the plane stranded in the desert and suggested that an inquiry would be held.

The Daily Guardian

NET SALES, 169,613

"Smith's Weekly" Net Sales Exceed 206,000 Each Issue.

FOUNDED BY JOYNTON SMITH

Vol. VI, No. 257 — Registered at the G.P.O., Sydney, for transmission by post as a newspaper — SATURDAY, APRIL 27, 1929

TO-DAY'S WEATHER: Cool and cloudy. (Special City Forecast.)   1D

Telephones: Editorial and Business — Publishing

AEROPLANE INQUIRY LIKELY TO BE HELD IN PUBLIC

PRIME MINISTER GIVES HINT: NATIONAL INTEREST

QANTAS PLANE ATALANTA (PILOT BRAIN) FLIES BACK TO BRISBANE IN RECORD TIME

LAND PARTY NOW ONLY TEN MILES FROM KOOKABURRA; AIR FORCE ACTIVITY

CAPTAIN HOLDEN'S REPORT MADE FOR AIR BOARD

A message received from Wave Hill late last night stated that the ground party being led by Lieut. Eaton was only ten miles north-west of the Kookaburra. Planes made two trips to direct them and dropped supplies.

Prime Minister Bruce said yesterday that he would be surprised if the Inquiry into the Southern Cross and Kookaburra were not open to the public. The personnel and terms would be announced soon, he said.

Stating that his report would be handed to the Air Board, Captain Les Holden refused to be interviewed when the Canberra arrived at Charleville yesterday from Longreach.

The Canberra refuelled at Charleville, and reached Bourke. It will arrive at Mascot Aerodrome about noon to-day.

Pilot Brain arrived at Brisbane yesterday at 4.10 p.m., after a trip lasting 4 hours 10 minutes. Thus, he beat by 10 minutes the late Keith Anderson's Sydney-Brisbane record.

Where Keith Anderson lies by the Kookaburra, in loneliest Australia. drew this picture from a description, true to detail, given him by Pilot Brain, of the plane Atalanta, which found the missing Sydney monoplane. On the right of the picture, near the machine's left wing, is seen the hole dug for water.

WAVE HILL, Friday.

TWO planes which made the flight to-day to locate the ground party were in charge of Flying Officer Ryan.

Ryan directed the party to the last waterhole, which is one and a half miles from their present position and 35 miles from Wave Hill.

The party has abandoned the car in which it was travelling, and is now proceeding to Anderson's plane with pack horses.

Ryan dropped further fuel supplies this afternoon and gave the direction of Anderson's machine.

MESSAGE TO AIR BOARD

LAND PARTY SHOULD REACH KOOKABURRA TO-DAY

MELBOURNE, Friday. — Following message was received late to-night by the Secretary of the Air Board, Major —

Fourth Wave Hill patrol, Flying Officers Ryan and Gerrand left Wave Hill at 8.50 a.m. to-day and returned at 10.5 having located party half a mile beyond car. Ascertained party was about —

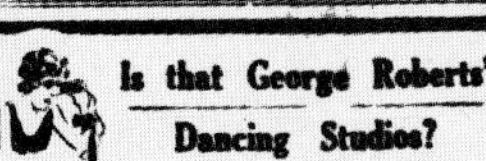

FRIENDLY BROLGA

Mr. Acton, manager of a station at Brunette Downs, whence the Atalanta left on her search flight for the Kookaburra, is on friendly terms with the Brolga — something between an emu and a native companion.

or this, or they would have left traces of their visits to the spot, which would have been sure to have attracted the attention of the land party.

The mystery of Hitchcock still remains unsolved, but the land party led by the black trackers should have little difficulty in finding him.

It will be difficult, it is thought, for the land party, when it reaches the Kookaburra, to send messages to Flying Officer Ryan.

FLIGHT INQUIRY REFERENCES

BRUCE INTIMATES IT WILL BE OPEN TO PUBLIC

CANBERRA, Friday. — The Prime Minister stated to-day that the personnel and terms of reference in connection with the Southern Cross and the Kookaburra were practically complete, and that an announcement would be made very shortly.

The question as to whether the inquiry would be an open one was, to an extent, the concern of the body appointed to make it, but in view of the fact that the public were so anxious to learn all the circumstances of the occurrence, he would be very much surprised to hear that it would not be open.

He stated also that he had received a wire from West Australia with reference to the stopping of Mrs. Hitchcock's pension, and was having inquiries made into the circumstances.

HOLDEN TO-DAY

NO TROUBLE WITH ENGINE; ONLY REFILLED WITH OIL

Mr. Eager stated last night that Captain Holden, flying the Canberra, expected to reach Mascot at 2 p.m. to-day. Of course this will depend on the weather.

Captain Holden reported last night that he had not experienced any trouble with his engine, as has been reported.

He had never any occasion to worry about it. All he did was to drain off the oil and refill at Longreach.

RECORD DASH TO QUEENSLAND

BRAIN'S HOP TO BRISBANE CREATES NEW RECORD

Mascot has seen no smarter take-off than that of the Atalanta at noon yesterday. And Brisbane no faster arrival.

The plane rose steeply after a remarkably short run, and without delay headed north.

In a few minutes it was out of sight. Brisbane is the miles away by airtime, but Pilot Brain expected to be at his home aerodrome there in time for dinner. He got in at 4.10 p.m.

The Sydney-Brisbane aero record was held by Keith Anderson.

In a Ryan monoplane he did the trip to the 20min. Brain beat that by ten minutes.

Atalanta used Shell oil and spirit on its big flight.

SIX NEW PLANES FOR AIR FORCE

WAPITIS AT RICHMOND NEXT MONTH

MELBOURNE, Friday. — Six new Wapiti aeroplanes for the Defence Department arrived by the steamer Tarmanki at Port Melbourne to-day.

The new planes are contained in 16 cases in the holds of the vessel. The machines will be assembled at Laverton to-morrow, and will be flown to Richmond (N.S.W.) for the Air Force camp next month.

When the Southern Cross was returned just the Wm. Glasgow, Minister for Defence, stated that the R.A.A.F. planes did not have the range for such search operations, and added that in Central Australia had the Wapitis which should have arrived early in the year.

SHOT HIMSELF AT CITY HALL

PIANIST'S STRANGE ACTION CAUSED ALARM

To shoot himself, a pianist, for some unknown reason, chose the front floor landing of the administrative building of the Sydney City Hall.

His name was given as John Alfred Watson, 42, pianist, of Bright Street, Camperdown.

The body was removed to the Morgue.

HOW QANTAS TAKES ON A QUICK JOB

FOUND KOOKABURRA IN 65 MINUTES!

FEAT IN ORGANISING

Qantas, which is the Queensland and Northern Territory Air Service, can show the right way to many highly boomed flight organisers.

PILOT Lester J. Brain, who on Thursday brought the plane Atalanta from Longreach to Sydney with pictures to "The Daily Guardian," told how Qantas acted when called upon to join in the search for Keith Anderson.

The Kookaburra, with Anderson and Hitchcock aboard, was missing since Wednesday, April 10.

On Thursday, April 18, Qantas in Brisbane received a message from Sir William Glasgow, Minister for Defence, saying that the Ministry would like a Qantas machine to help in the search.

Lester Brain, who is manager of Qantas Brisbane branch, was at Eagle Farm (Brisbane). He got into touch with the managing director, Mr. Hudson Fysh, who was in Brisbane attending to the new aerial mail route to Charleville.

HE KNEW TERRITORY

Fysh told Brain that the best machine to use would be the Atalanta, 460 h.p. Bristol Jupiter engine, a D.H.50 machine, built by Qantas at Longreach (Q.). She is the fastest performing machine in Qantas for speed and climb.

Brain was selected as pilot because he knew the Territory area in which Anderson was lost. Four years ago he flew the locality looking for a suitable gold outcrop. His only acquaintance with the Atalanta had been a trial flight of 10 minutes.

Still on the same day, April 18, Brain rang Mr. Robinson, of the Queensland radio service.

"Can I have wireless on the Atalanta before daybreak?"

"Yes" said Robinson.

They chose Brunette as radio operator, having explained to him the duties of the trip.

The wireless, run by battery, allowed constant sending while in the air, with a receiving apparatus either on the ground.

WIRELESS STATION AT BRUNETTE DOWNS

Captain Brain left this wireless station at 7 a.m. local time on the memorable Sunday when his searching plane, Atalanta, found the lost Kookaburra, with the body of Keith Anderson lying in the shadow of her wing.

been found alive if the full search had been organised sooner?

The coming inquiry will look for an answer to that question.

The Southern Cross is at Narromine. It is due at Sydney to-day.

"MYTH" PLANES

FOR the names of its planes, Qantas drew on Greek mythology. Besides the Atalanta, the fleet includes the Hermes, the Iris, the Pegasus, the Perseus, and the Apollo.

NINE DAYS MISSING

The Atalanta crew lunched at Longreach at 3.55 p.m., and took off about 4.20 a.m., the provisions and equipment checked yesterday.

On Saturday April 20, starting at 4.20 a.m., the Atalanta flew past Cloncurry west, and on to Brunette Downs, her wireless working perfectly all the time. On Sunday, April 21, the fateful day of discovery, the Atalanta set out from Brunette Downs at 7 a.m., and reached Newcastle Waters at 8.50 a.m. Refuelling, she set out from Newcastle Waters at 11.5 a.m.

The rest of the complete story was written in "The Guardian" yesterday. Exactly 65 minutes later the Atalanta, at 12.10 p.m., sent out her radio that she had found the Kookaburra.

Would Anderson and Hitchcock have

CAPT. MASE STARTS ON LONG HOP TO NEW ZEALAND

LONDON, Friday. — Captain Frank Mase started at 6.15 a.m. to-day from Lympne, Kent, on his flight to New Zealand.

He proposes to fly first to Australia, and to hop off from Hobart to Christchurch.

GRAND DUKE MICHAEL DEAD

RUSSIAN EXILE ENDS HIS DAYS IN ENGLAND

LONDON, Friday. — The Grand Duke Michael, a member of the Russian Royal Family, died at his residence, Regent's Park, from influenza.

The news of the revolution and the confiscation of the Czar, Grand Duke Michael fled from Russia and settled in England.

Arrangements being finalised through the night, the crew of three had to —

JOLT FOR BIG BILL THOMPSON

"GOD SAVE THE KING" IN CHICAGO

CHICAGO, Friday. — "God save the King" we heard yesterday in the streets of Chicago, the city whose mayor has frequently criticised King George and Britain — and not a soul objected.

IMMEDIATE cause of this unusual happening was the contingent of Young Australia Lounge boys who arrived with the Union Jack and the Stars and stripes flying.

A town's mouse warnings had been received that if the boys went to Chicago there would be trouble, but nothing developed.

They marched through the long, mellow streets during the busiest hour of the day, scores of local school cadets.

THEN THE BANDS PLAYED

Band of the local boys played the French National Anthem, while the U.S. Light responded with "The Star Spangled Banner."

Down La Salle Street past the darkened offices of "Big Bill" Thompson, the Mayor, marched the youngsters.

The crowd showed nothing but cheerful goodwill towards the visitors.

PUT AUSTRALIA FIRST

BODALLA, Friday — At an Anzac Mass at St. Mary's Roman Catholic Church, Father John Roche said that as an Australian-born priest, he was proud of the Australian flag.

"Make this your motto," he advised: "Australia and the Australian flag first every time." If you do this, Australia will be proud of you.

R.A.F. AIRMEN FAIL IN NON-STOP FLIGHT RECORD

336 MILES SHORT OF ITALIAN RECORD

KARACHI LANDING

"GUARDIAN" SERVICE.

LONDON, Friday.

The British airmen who are flying the R.A.F. Fairey monoplane failed to break the world's non-stop flight record. They landed at Karachi at 11 p.m. after covering 4130 miles in 50 1/2 hours.

Another 336 miles was required to break the record.

PASSING Jask, Persia, they had almost 2000 miles between them and their destination (Rangoon, Southern India). Pilots are Squadron Leader Jones-Williams and Flight-Lieut. Jenkins.

FROM ENGLAND TO INDIA

With 1000 gallons of petrol, the plane left Cranwell, Lincolnshire, at 7.15 on Wednesday, and passed Basra over half-way, 12 hours later.

Distance non-stop record of 4418 miles was made last July by Ferrarin and Delprete, Italian airmen, who flew from Rome down to the sea, ran east, across the Atlantic, to Brazil.

The British airmen are out to beat this record, and the one for duration of the air held by German fliers who remained aloft without refuelling for 65 hours.

£50,000 JEWEL ROBBERY

LONDON, Friday — Jewellery worth £15,000 including a pearl necklace valued at £12,000 was stolen from a fashionable West End hotel.

The valuables belong to Mrs. Anna Jurgens, wife of the "Margarine Millionaire."

knee, which handicapped his movements for some days.

Since they had no way of indicating their reason for not proceeding, the men attached a cord between two poles, with weighted message bags at each end. This practice was fairly common, as a skilful pilot could pick up the cord with his tail-skid.

Around 2 pm, two DH 9As appeared and dropped a message which gave the direction and distance to the *Kookaburra*. In return, the ground party pointed to the message cord. Ryan made a run at it, but misjudged his height and bumped the ground with his tail-skid. He managed to keep control, then prepared for a second attempt. Realizing the danger involved, Eaton signalled him away, but Ryan ignored the frantic waving and came down a second time, scooping up the cord. The disadvantage of this primitive method was that Ryan could not read the message until he landed at Wave Hill.

Sambo and Daylight returned with the horses and reported that their companion, who had been guarding the Buick, was well and had been delighted to see them after his lonely vigil.

Next morning Sambo remained at the blow-hole to care for some of the horses, while Daylight accompanied the other men and a few horses, to find the *Kookaburra*. At 11 am a DH 9A appeared and pointed out the necessary directions, but the journey was still a nightmare. Eaton's diary recorded:

'As the horses pushed through the scrub, myriads of flies rose and settled on horses and men alike. They would not be brushed off and we were just covered in a sticky mess. Twenty-four hours passed and Moray, an experienced and good bushman, thought we were at the end of our tether. However, my aeroplanes from Wave Hill, which had been guiding us, dropped a message saying we were only 1½ miles* from the *Kookaburra*. On we went and made it.

'Time was against us. We found poor Hitchcock lying under the wing dead. We could not see Anderson and, as we buried Hitchcock where he lay, the Aboriginals searched. It took them nearly an hour to track Anderson who, at the end, had wandered away, divesting himself of clothing, and perished within 500 yards** of the forced landing. We buried Anderson where he lay. The Lord's Prayer was said as the airmen were laid to rest.'

In order to bury Hitchcock, Eaton turned the plane around, so that it faced the bush, whereas

The *Kookaburra* with the rough grave where Hitchcock was buried, under the plane's wing.

* *2.4 kilometres.*

** *457 metres.*

they had found it pointing in the direction of the cleared runway.

The ground party discovered that Lester Brain had been incorrect in his assumption of the identity of the dead man whom he had seen from the air. The man under the wing was Hitchcock, and they identified him by the bandages on his forearm. The nurses at Alice Springs Hospital had told Eaton that Hitchcock was suffering from slight blood poisoning in one arm and they had treated it prior to his departure.

The ground party also discovered that the hole under the aircraft wing had probably been dug in an attempt to find water. Daylight explained that his tracking of Anderson indicated that the dead aviator had walked to the south-west for about 455 metres, with two water bottles and a cushion. He then sat down under some scrub and, some time later, had proceeded with his search.

By that time 'the whiteman was very sick', said Daylight. His tracks led through the scrub, instead of around it as previously. A little further on he had fallen, but continued by crawling. His tracks then started to circle to the right, indicating that Anderson had lost his sense of direction. Some distance on, Anderson discarded his coat, flying helmet and goggles and soon afterwards abandoned the water bottles and cushion. The track continued to circle to the right and Daylight found a pile of personal possessions consisting of a shirt, aircraft log, watch and scarf. Forty-five metres further on Anderson had inhaled his last breath.

Anderson did not of course foresee his own tragic end. He had started to record the events but eventually he gave up, no doubt weakened by thirst. On the fabric of the *Kookaburra's* rudder, he wrote:

'DIARY 10/4/29 to -/4/29. Force landed here 2.35 pm 10th April 1929 thru push-rod loosening No. 2 cylinder cutting out (as at Algebuckina SA on 9/4/29 but temporarily fixed K.V.A.) exhaust valve and 25% h.p. Cleared bit of a runway here which turned out just insufficient or engine coincidentally lost power. Since 12/4/1929 all efforts of course same next to nil, thru having no water to drink, except solutions of urine (with oil, petrol, methylate [spirits] from compass) directed on obtaining sufficient power from engine to permit successful take-off. No take-off able to be attempted since 11/4/29 due increased debility from thirst, heat, flies and dust. Left Stuart (Alice Springs) 7.15 am local time and followed

Eaton and Moray, the pastoral supervisor of Wave Hill station, stood beside the *Kookaburra* after they had buried Hitchcock's body.

Sergeant Douglas and Flight Lieutenant Eaton beside Anderson's grave.

Right, Anderson wrote his diary on the fabric of the *Kookaburra's* tail, which Charles Eaton cut out to take back to the Air Board. The diary is now in the West Australian museum in Perth.

telegraph line for 100 miles* which was intention. Cut off then direct for point between Wave Hill and Ord River Downs. On a/c cross winds and inaccurate compass and having practically only sun for guidance as large map showed only featureless desert determined to above or nor'rard of course which am sure have done. As was in the air 7 hours and am. Pretty confident had 'duckpond' on my starboard. I figure position now to be ...'

In his most dramatic comment on the unhappy incident, Eaton said that 'we saw it, high up, right above the *Kookaburra,* in an otherwise clear sky, a perfect cross formed by what may have been a cirrus cloud. We did not say much to each other, but I think we all thought a lot as we made our weary journey home.'

Whilst the ground party searched the grim site, the two DH 9As dropped a couple of wreaths and returned to Wave Hill.

There have been mischievous reports that Anderson finally ended his own life with a pistol. Stan Cawood, a member of a subsequent expedition to recover the bodies, claimed on a

Below, the *Kookaburra's* tail rudder showing the remaining hole after the diary was cut out.

* *160 kilometres.*

CH-73/61

television programme that a discharged gun had been found at the spot but 'it was agreed not to say anything for fear of causing distress and trouble'.

Eaton completely refuted this claim by stating emphatically that when they found the body there was no sign of injury and certainly not a gun wound. He said that his party had not seen a weapon at the scene.

Douglas' account also rejects the suicide rumours, paying tribute to a brave man:
'It was an extremely sad moment for us, as we realized the agony of mind and torture of body he must have gone through. It was also clear that he kept his mind to the last and fought against the odds and had died fighting. His courage and determination must have been unlimited.'

Eaton abandoned his plan to carry the bodies back to Wave Hill because the horses had been without water for too long. Similarly, they did not attempt to fly the *Kookaburra*. The plane itself was in reasonable condition and Douglas

quickly had the engine running. They found 90 litres* of fuel in the tank and enough engine oil for about three hours of flight. However, one tyre was flat and several holes in the fuselage fabric would need to be repaired. This work, together with the necessary extension of the runway, could have taken a further day.

Eaton was keen to try and take off in the *Kookaburra*, but Moray insisted that because of the water shortage, they should leave the site by 6 pm that day. They made a further inspection of the plane and found some food still in the cockpit; probably the lost men were so thirsty after trying to clear a runway through the scrub that they could not eat.

Douglas removed the joy stick** (control-lever of the plane), presumably to prevent anyone else flying the *Kookaburra* without permission. Eaton cut from the tail, the fabric bearing Anderson's log , to give to the Air Board. Douglas took some photographs of the desolate scene and, at 5.30 pm, the party left.

Daylight, the stockman, was asked to lead the party during the night, to reach his companion at the blow-hole and again he showed his

*This was an estimate. The amount was later found to be about 140 litres.

**The joy stick was later given to Bon Hilliard and is now in the old people's home at Alice Springs.

Sergeant Douglas (left) and Flight Lieutenant Eaton arrive back at Laverton, Victoria, after completing their assignment and burying the dead aviators. Only two of the five DH 9As returned to base, but none of the 10 RAAF men was injured.

intuitive skill. Douglas' diary records:
'I was at a loss to discover how Daylight managed to lead us in a direct line through the scrub without once faltering or deviating from his course. Without his assistance, we could not have risked the night ride, even with Mr Moray's compass. Sambo was not the least bit excited by our arrival. He acted just as if he expected us to show up when we did.'

Next morning, the party left the blow-hole and reached the billabong soon after lunch. By then the water was drying up rapidly and some of the horses had existed without a drink for 52 hours while covering about 144 kilometres. The horses were refreshed in the most adequate manner with the failing water supply. The following day, Daylight continued his amazing performance by leading them straight to the abandoned Buick, where the third Aboriginal was patiently awaiting their return.

The radiator was fixed and refitted and the flat tyres, changed. The stockmen were instructed to lead the horses back to the homestead at a leisurely pace, while the others went ahead in the car.

Douglas became ill, so he and the other men from the RAAF managed a few days rest before leaving. Even now, their problems were not over. When Douglas attempted to start his aircraft's engine, it caught fire and was thus destroyed. According to historian John Haslett, the remains of the engine and the DH 9A airframe are at Darwin's aviation museum, and the propeller adorns a wall of the recreation room at the new Wave Hill homestead.

With Douglas' machine out of action, only two of the DH 9As were serviceable. Eaton's had been wrecked during the crash landing, and another was left at Tennant Creek because of repeated engine trouble. So Eaton[*] and three of his colleagues climbed into the remaining DH 9As and flew to Melbourne, arriving on Saturday May 11th, 1929, after four extremely strenuous weeks.

* *Sadly, Flight Lieutenant Eaton died on November 12th, 1979.*

The Thornycroft expedition

Once the *Kookaburra* was found, the responsibility for the bodies of the dead aviators passed into the hands of the Minister for Home Affairs and Territories, Mr Abbott. The epitome of a politician who refuses to be desk-bound, he took a direct and personal interest in the affair. Almost immediately after the events described in this chapter, he chartered the airliner *Canberra* piloted by Captain Les Holden, the man who had found the missing *Southern Cross* in the Coffee Royal Affair. In a two-week whirlwind tour, Mr Abbott flew 10,000 kilometres over the inland, returning to strongly reaffirm his faith in air travel. After making glowing references to Captain Holden, he said: 'At Newcastle Waters, I told Captain Holden I would like to see where the *Kookaburra* lay. It was about 40 miles* away in dense country, but Holden replied, "I'll see if I can find it", and a half hour later we were on the spot.'

When the Prime Minister decided that the Commonwealth would reclaim the bodies, Mr Abbott sent word to relatives and the Press. He was immediately contacted by Mr J.O. Johnson, a director of Thornycroft Australia, the importers of the British truck of that name. Mr Johnson offered to make available — at no cost to the Commonwealth — a brand new A3 Thornycroft truck to help in the recovery mission. His only stipulation was that the Commonwealth should meet all expenses, including the cost of the personnel supplied with the truck.

The A3 Thornycroft was rather a novelty, since it had four-wheel drive controlled by a special gearbox including twin rear axles. Like a modern four-wheel-drive unit, it had a special low driving range for use in difficult terrain.

Mr Abbott accepted the loan of the truck and arranged for the railway system in South Australia to transport it — on a special flat top behind the main carriages, to the rail-head south of Alice Springs.

The plan was for the Thornycroft truck party to operate from Alice Springs, via Wave Hill. The three-tonne truck would be laden with 2,730 litres of water, petrol and oil, in order to make the overland trip from the track to the plane, and also to drop off water and fuel supplies on the way to be used for the return trip.

The party, according to the plan, would also inspect the *Kookaburra* and, if practical, tow it to Wave Hill. Mr Abbott announced that Flight Lieutenant Eaton would be asked to return from Point Cook near Laverton, to serve as an aerial guide but, for some reason, this did not eventuate. The truck party received no support from the air.

The men chosen to join the truck party included Frank Nottle, Thornycroft's South Australian representative, Les Miles (a driver/mechanic) from Sydney, William Berg (a writer/photographer from the *Guardian,* a Sydney newspaper) and Mounted Constable George Murray.

An Adelaide firm, Pengelly & Knabe, undertook to provide special coffins for the occasions, consisting of pine boxes sealed in special lead casings, with ornamental outer caskets.

William Berg and Les Miles wrote their own diaries of the journey but, in places, the wording of Les Miles' journal closely follows that of Berg's newspaper articles so the journal may have been written after the expedition was over.

Both men lived in Sydney and travelled by train to Adelaide, where they arrived on May 23rd, 1929. They were briefed at the office of the Commonwealth Director of Works, where they met Frank Nottle. The truck was modified for its overland task; the front mudguards were removed to allow easier passage through dense scrub and the bonnet cowling was taken to maximize the cooling of the engine. A massive bumper was fitted because the truck would be forcing its way, where necessary, through the scrub. A mesh screen was placed across the radiator to keep out insects.

The Thornycroft truck was apparently not given a nickname, although this was a customary practice at the time. It left Adelaide by train on Friday May 24th, and the men followed on Monday. At Terowie, 224 kilometres from Adelaide, the standard-gauge line ended and the men and truck were transferred to special carriages on the narrow-gauge running to the inland. They reached the rail-head near Alice Springs on Thursday May 30th, where they met Constable Murray, who had been appointed by the Commonwealth as expedition leader. Constable Murray arrived in his Model T Ford and told the Thornycroft men that he would be driving it overland, to accompany the truck.

---

* *64 kilometres, but the distance was in fact 160 kilometres.*

On its way from Alice Springs to the location of the *Kookaburra,* the six-wheeled Thornycroft truck initially travelled along stock routes and followed overland telegraph lines. This photograph, taken between Banka and Powell's Creek, gives an idea of the height of the turpentine scrub in which the *Kookaburra* was forced to land.

As outlined by the Commonwealth Home and Territories Department, the original plan called for the truck party to go to Wave Hill and follow the horses' tracks made by Eaton's ground party. This involved hundreds of kilometres of extra travel, so Constable Murray, supported by Mr Cawood, (the local Government Resident), proposed a shorter route. This was finally accepted by the Government, on the condition that Murray and Cawood would accept full responsibility.

In his diary, Berg graphically described the new route:

'A hoof mark in the centre of a continent ... this was the target of the Thornycroft expedition. Somewhere in the wilderness were the horses' tracks of Lieutenant Eaton's party ... to follow the trail would have meant delay, perhaps failure. The expedition therefore took the risk of a short-cut from Newcastle Waters, trusting to strike the trail taken weeks before by Eaton's party. We owed our success, perhaps our lives, to the fact that we had with us the greatest pathfinders in the world — the Australian black fellow.

'At Newcastle Waters, the north-most point of our route, we picked up four Aboriginal guides. Perhaps in the days of their glory they would have been king, priest, hunter and warrior. Today, they are just Bob, Tommy, Hughie and Jack, humble servants of the white man.'

Stan Cawood, the son of the Government Resident, had volunteered to join the party as a cook, so the eventual expedition consisted of five white men and four Aboriginals. The truck aroused almost as much interest as the expedition, as six-wheel-drive machines were then almost unknown in the inland. Some landholders travelled for great distances to see it and assess its potential for carrying heavy loads over difficult terrain.

The day after reaching Alice Springs, the expedition set out on the 815-kilometre trip through central Australia to Newcastle Waters*. They experienced a fair road to Connor's Well, then travelled through mulga country to Ti-Tree Well, which they reached on Saturday evening. From the start, they found the constant

* *The actual distance from Alice Springs to Newcastle Waters along the Stuart Highway which was completed in 1941, is 788 kilometres.*

buzzing of flies a greater inconvenience than the rough ride. The fly menace reached its zenith at Ti-Tree, where in order to eat a camp meal, they found it necessary to light a fire made from camel dung and sit in the smoke to eat their food. 'I don't know which was worst — the flies or the smoke!' wrote Miles.

Next morning the group saw — in the distance — Mount Stewart, which was said to be the exact centre of the continent. Just north of Barrow Creek they passed the 'Devil's Marbles'; huge red rocks with sides so slippery that it was impossible to climb them. As others have done before and since, the men wondered how some of the massive 'marbles' had come to balance on top of much smaller ones.

On Sunday evening, the men camped south of Tennant Creek, having covered 512 kilometres in two days. By this time, they were suffering from dysentery, no doubt caused by the water they drank. Nottle was especially ill, but this did not prevent an early start on Monday. Within minutes of leaving, they came across the DH 9A which Eaton had wrecked in his forced landing while searching for the *Kookaburra*.

Right, Les Miles, the Thornycroft driver, beside an ant hill.

Tuesday took them to Bonning Station where they enjoyed morning tea with Mrs Bonning, the second white woman they had seen since leaving Alice Springs, then 640 kilometres away. Both the truck and Model T were running faultlessly but on Tuesday night, only 100 kilometres from Newcastle Waters, Constable Murray crashed into a wash-away, ran off the road and ended with his trusty Ford balanced on its radiator in a ditch. The Thornycroft had little trouble pulling it out and, despite a bent axle, the car hobbled to Newcastle Waters where Charles Berg, the local mechanic, repaired it. The journey from Alice Springs had taken five days.

The Aboriginal guides joined the expedition at Newcastle Waters. The men also loaded on so much fuel, water, oil and supplies — not to mention the heavy caterpillar tracks which they were carrying in case the truck became bogged — that the big Thornycroft carried a total load of three and a half tonnes, plus the crew.

Anxious to lose as little time as possible, Constable Murray ordered an early start for Thursday morning. He led the way in his Model T, but unfortunately the engine broke down with ignition trouble and the party returned to Newcastle Waters for new parts. When these were fitted, they set out once again on the trip to the site where the *Kookaburra* lay.

Murray followed the Murranji stock route for about 60 kilometres, then turned to follow a railway survey line to the south-west. The marker pegs were so overgrown with scrub that it took the combined skill of the four trackers to keep them on the line. The truck made slow progress through the loose sand and it is doubtful if a conventional two-wheel-drive machine could have continued. The Thornycroft rumbled on, hour after hour, never out of first gear, but neither protesting nor overheating. The wheels were fitted with high-pressure tyres, then relatively new, as many punctures resulted from the sharp turpentine scrub. The detachable rim system of the wheels made it easy to repair them and the truck carried an engine-driven air pump which proved to be of great benefit, as no one had wanted to inflate the tyres by hand in the torrid heat.

In places the scrub was as tall as the men and Nottle experienced difficulties in finding his way through it without hitting ant hills or tree stumps. Since the Model T was following the path flattened by the truck, it had less trouble in getting through.

The first night out from Newcastle Waters they camped on the edge of the bullwaddy country, where gnarled boughs erupted from the red sand in grotesque shapes and the bush was so thick that the men despaired of penetrating it. All of the next day was also spent fighting the undergrowth. The technique was to find the least dense bush and charge into it, but this did not always work as they frequently came across their own tracks. For five days, the battle raged, with the Aboriginal 'boys' scouting ahead of the truck, trying to follow the survey lines. Suddenly, on June 11th, buried in the spinifex, they stumbled across horses' tracks which had been made by Eaton's party. Berg recorded that the tracks were so faint that they were invisible to a white man's eyes and it made him 'go cold to think what might have happened if the blacks had not been with us'.

Berg and the others were well aware of the danger of being lost. They were also anxious about their fuel supplies, even though they had laid 'dumps' behind them. To make matters worse, after 30 kilometres of exceptionally difficult progress, they discovered that they were on the wrong track. As Miles reported in his diary:

'We were on the move at 7 o'clock retracing our track back for nine miles* where we found Eaton's horses had branched off in a N-W direction. Here Constable Murray had a lot of trouble with the trackers, they wanted to desert us, Murray had to take the drastic measure of threatening them with his revolver before they would continue on. They were far from friendly for the rest of the day and we had to watch them closely in case they turned on us.

'We followed the horse tracks in a N-W direction for about a mile and came to the blow-hole mentioned in Lieutenant Eaton's report. We also found letters from Ryan (the pilot of the plane) which explained the dead-end horse tracks of the previous day. He (Ryan) had been delayed at Newcastle Waters with a leak in the radiator of his plane and Lieutenant Eaton had gone on without guidance of the planes and had got lost.'

The wasted time and petrol distressed the Thornycroft party, as they had travelled an extra 30 kilometres each way, over some of the worst terrain they had encountered. They eventually found the tracks leading to the *Kookaburra*, but it took a further day to push through the heavy scrub.

'Suddenly the wall thinned and was gone,' said Berg's diary. 'And we saw before us a vast desert of black ashes and red sand, on which stood row upon unending row of charred and naked skeletons of turpentine and turkey bush. As far as the eye could see, nothing broke the monotony of this infernal landscape.'

The *Kookaburra* was nowhere in sight. Grouped into pairs, a white with a black, the party went searching, but still the plane could not be found. The men returned to the truck and went charging through the blackened bush in a desperate attempt to find it before the growing fuel shortage forced a retreat.

At 4 pm — eight days after leaving Newcastle Waters — the party burst upon the scene which Berg described as the 'loneliest grave a hero has ever known'.

The Aboriginals were riding on top of the truck and were first to spot the *Kookaburra*, its silvery wings glittering in the sun. Berg reported:

'We had a feeling of triumph at having achieved

* *14.5 kilometres.*

Above right, Bob, Tommy, Hughie and Jack — the four stockmen accompanying the Thornycroft party.

Right, Frank Nottle, Les Miles, Stan Cawood and Constable Murray in front of the *Kookaburra* at the scene of the tragedy. The fifth white man in the group — Charles Berg — was the official photographer and reporter.

The official records of the Thornycroft expedition scarcely mention the Model T Ford, (owned by Constable Murray), which accompanied them to the *Kookaburra*. It is doubtful that the car would have made the distance if it had not been able to follow the path which was often bulldozed through the scrub by the big six-wheeled truck. Here the Model T Ford is shown with the truck near the *Kookaburra*.

Above right, this view of the *Kookaburra* was taken from Anderson's first grave. The tracks of the truck are visible.

Right, the truck party folded back the *Kookaburra's* wings to prevent possible damage in strong winds. This was the last photograph taken by Charles Berg before the Thornycroft party left the scene of the tragedy. Hitchcock's grave is marked by a stake and Constable Murray is sitting on the wheel.

what men who knew the country had declared almost impossible, but this was swallowed up by the awful tragedy and desolation of the scene.

'In a few moments, the Thornycroft (halted) alongside the mono-plane. She stood facing the east, tilted to the south by the depression of a punctured tyre on that side. Somehow the beauty and grace of this man-made machine struck us poignantly as we saw her there, marooned in the pitiless waste of primeval nature.'

Constable Murray had the men exhume Hitchcock's body, which had been buried under the plane's right wing. They dug a trench parallel to the grave, laid down matting and rolled the body onto it. Four men lifted the matting by the corners and lowered it into the pine coffin. The leaden overlay was sealed. Berg photographed the sad scene in the fading light.

Whilst the white men exhumed the body, the Aboriginals were sent to locate Anderson's grave, which was 365 metres from the plane and marked with a rough pole cut from nearby scrub.

By 7 pm, Hitchcock's casket had been sealed but the Aboriginals, who had not previously known the nature of the search, were now terrified by the exhumation. That night they slept between two guardian fires, close to the white men, although it had been their practice to camp some distance away. Tortured by superstition, they became increasingly agitated when a cool wind rocked the plane, causing eerie creaks in the half-light. Early in the morning, two white men arose and folded back the wings in case the wind toppled the plane over.

At dawn, Murray and Berg rose and noticed a strange blue light had descended on the vegetation. They walked quickly to Anderson's grave and exhumed his body. By the time the others had arrived with the truck, the body was ready to be sealed in its coffin. Berg investigated the area and found a map and an envelope containing money, which he handed to Murray. The constable conducted his own search and found a cuff-link belonging to Hitchcock.

Murray had decided that it would be impossible to tow the plane back to Newcastle Waters as planned. However, since it may be flown out at a later date, he instructed the Aboriginals to extend the runway, using picks and shovels, whilst the body was being exhumed. The white men later joined the work and, after four hours, they had created a runway, 53 metres wide and 293 metres long.

Berg said that he had counted 11 separate wheel impressions in the rough runway that Anderson and Hitchcock had scratched out with their hands, suggesting that five or six attempts had been made to fly the *Kookaburra*. Further, he said that they had found one punctured tube but a thorough search did not reveal a repair kit. This, he believed, was the final blow which cost the aviators their lives.

With his mind now on the return journey, Murray was particularly concerned about the fuel supply. Even counting the dumps they had laid along the way, it was doubtful that they could reach Newcastle Waters with fuel to spare. If they became lost again it would be almost certain that they would run out so, as an emergency precaution, they drained the 140 litres of petrol which remained in the *Kookaburra*. They then departed for Newcastle Waters.

The trip from Newcastle Waters had taken eight days to cover a distance of 470 kilometres*.

** The actual distance was closer to 160 kilometres.*

By the time they had reached the stranded plane, the Thornycroft men were short of petrol. Driver Les Miles syphoned about 140 litres from the *Kookaburra* to replenish the truck's supply. Presumably his cigarette was not lit.

As a result of the tracks and notes they had made on the way, the party completed the return trip in two and a half days. They needed to drive only 312 kilometres as this time, they did not lose their way. One reason for their haste was that they were two days overdue at Newcastle Waters and they feared that a search might be launched.

The first day after leaving the *Kookaburra*, they covered 50 kilometres but the following day they travelled for 11 hours and covered 100 kilometres, averaging three and a half miles per gallon of fuel, (80 litres per 100 kilometres). The heavy fuel consumption was caused by the tough terrain which kept the Thornycroft in first or second gear all the way. They managed 163 kilometres in 12 hours on the third day, partly because they had reached the Murranji track which, for the first time in 10 days, allowed them the luxury of travelling in top gear. They reached Newcastle Waters at 6.30 pm that day.

Their haste proved costly, since the tyres were punctured often. Once they collected four flat tyres in two hours. They also became in-

creasingly concerned about the fuel situation. On one occasion, they had used all the loose cans and the truck's tank was down to the last 27 litres when they reached their next fuel dump.

The Aboriginals began to anticipate the return to their homes at Newcastle Waters, where they could regale their fellows no doubt, with horrifying stories about white men disturbing graves.

After a good night's rest, the white men set out for Alice Springs, with Murray leading the way in his Model T. Abruptly, at Tennant Creek, he left the party, saying that he wanted to make a 'call down the road'. He rejoined them at Barrow Creek, having captured an alleged murderer named Willoberta Jack. He also apprehended Jack's woman.

Willoberta Jack was unceremoniously chained to the vehicle since the constable regarded him as a prize capture and had no intention of losing him. Jack was wanted for the alleged murder of a white settler named Harry Henty. For six months he had evaded capture by moving from hill to hill with his woman, always a step ahead. To guard his captives at night, Murray placed steel rings around their

A page from the diary which was compiled by Les Miles, a member of the expedition. It was probably written after they had returned.

Murray & Berg went to Anderson's grave, & by the time Nottle & myself had finished off Hitchcock's coffin & arrived with the lorry, they had Anderson's body ready for its transfer to the coffin. In both cases we dug a trench alongside the body & placed matting down, the body was then worked on to the matting, four of us taking a corner of the matting each would then lift the matting & body, & place it into the casket.

While the exhuming of the bodies was carried out, the blacks were given axes, picks & shovels, & set to work clearing a runaway, as we had decided it would be impossible for us to remove the plane without severely damaging it, and, as a result, it must be flown off later.

This runway the whites improved later, & finally it extended 320 yards in front of the plane, & was 60 yards wide, running N. & S. to suit the prevailing winds.

A close inspection of the plane showed it to be in excellent condition except for several vents in the fabric, caused, apparently, when the plane landed, or in the endeavors to take off again.

Extending north for about 150 yards from where the plane lay was a roughly hewn takeoff through the scrub, made by the unfortunate men in their efforts to save themselves. And what a task it was!

On his way back from the *Kookaburra*, Constable Murray captured an alleged murderer named Willoberta Jack. He kept him chained to the vehicle until the truck party reached Alice Springs.

necks and chained them to the vehicle or a stout tree.

His vigilance in capturing them was applauded at the highest level. The Minister for Home Affairs and Territories, Mr Abbott, told the Press:

'Constable Murray heard two blacks talking about the alleged murder of a white settler named Henty, who was supposed to have been shot by a native. On the strength of what he had heard, he left his party, walked through the night to a native camp, held the suspect at revolver point and made the man's lubra handcuff him.' Abbott went on to say there were only 12 constables to maintain law and order in the vast territory.

'Their motto is "Get Your Man",' he said, 'and like the famous north-west mounted police of Canada, they lived up to it.'

The Thornycroft truck party made good progress to Alice Springs where flags were at half mast and men, women and children lined the main street, bare headed, in a silent tribute to the dead aviators. A gasp greeted the arrival of Constable Murray who paraded Willoberta Jack and his woman in handcuffs.

For the first time in weeks, the white men relaxed; their gruelling 1,500-kilometre trip was over. After a celebration dinner, they boarded the train with the coffins, to return to Adelaide.

It is believed that for the next 32 years nobody sighted the *Kookaburra* again. There have been reports that the plane had been towed from its original landing site, but this seems unlikely. Lester Brain was told in 1929, that a group of young people from Newcastle Waters who wanted to learn to fly, had towed the plane for some distance before abandoning it.

Researcher Dick Smith is certain that this did not happen. He believes that the story eventuated because the Thornycroft truck party had announced that they would tow the plane back to civilization. When Dick Smith's group returned to the *Kookaburra* site in 1978, they searched the area and found enough debris from the plane to confirm that the *Kookaburra* was still where Anderson and Hitchcock had left it. And that is where it stayed, despite pleas from Anderson's friends and relatives that it should be recovered.

At one time, Bon Hilliard's father announced that if the Government did not initiate its recovery, he would reclaim the plane himself. There was also an offer from Charles Berg, the mechanic who had fixed the Model T at Newcastle Waters. He said that he would tow the *Kookaburra* to Newcastle Waters and deliver it to Darwin or purchase the plane himself. No one offered the necessary financial backing, so the *Kookaburra* remained in the dismal grasp of the desert; a solitary reminder of the sacrifice made by Anderson and Hitchcock.

History does not record the fate of Willoberta Jack. His capture in such unusual circumstances must be listed among Australia's most extraordinary murder cases. It was another bizarre link in the chain of events which commenced the day that Kingsford Smith and Charles Ulm decided to fly to England.

A page from the *Daily Guardian* showed some of William Berg's photographs of the Thornycroft truck party at the site of the *Kookaburra*. In the story he stated that the burial experience was 'worse than the war'.

"GUARDIAN" PHOTOGRAPHER'S STORY OF KOOKABURRA EXPEDITION

Photographer W. N. Berg's graphic pictures tell the story of the burial expedition to the Kookaburra, Mr. Berg having been a member of the expedition. (1) Constable Murray, leader of the expedition, takes a compass reading. (2) The black boys, who materially assisted the party, pose in front of the plane. Left to right: Bob, Tommy, Hughie, and Jack. (3) The black trackers showed uncanny skill in following the horse tracks left by Eaton's party. (4) Left to right: F. Nottle, L. Mile, S. Cawood, and Constable G. Murray, taken at the scene of the tragedy. (5) S. Cawood, a member of the party, scales one of the few tall trees encountered en route, in an endeavour to pick up some landmark.

S. CROSS HOPS OFF

FLIGHT TO SINGAPORE COMMENCES

GOOD PROGRESS

Hopping off at 5.10 p.m. (Sydney time) from Derby for England, the Southern Cross commenced the long flight across the Indian Ocean to Singapore. The distance is approximately 2050 miles.

A BROKEN distributor drive pinion of one magneto on the port engine delayed the take-off slightly, but within less than two yards the plane headed across the Straits between Java and Sumatra.

The following messages detailing the plane's progress were received in Sydney during the night.

(West Australian standard time is two hours ahead.)

5.25 p.m.—Off the coast five miles south of Cape Leveque, crossed 300 south, long. 121.12 east. Speed 109 knots. No more now; see you later.

10.1 p.m.—Dead reckoning position Lat. 15.12 S., Long. 120.170, speed 96 knots, altitude 3000 feet. Few bumps but all O.K.

11.2 p.m.: Dead reckoning position, lat. 14.20 south, long. 118.45; speed 95 knots; altitude, 1300 feet. Not very cold. Nice flying condition at present.

11.32 p.m.: Latch, and I have been touching up the good food—Smith.

12.19 a.m.: Here 19 p.m., dead reckoning. Position, longitude 117.27 east, course 206 deg., approx. Speed, 94 knots. Altitude, 3600 feet. All O.K.

3 a.m.: Here 3 a.m., dead reckoning—latitude 11 deg. 10 min. south; longitude 112 deg. 48 min. east. Course, 301 deg. Speed, 97 knots. Altitude, 2000ft.

SOURABAYA MESSAGE

At 9.45 p.m. the following message was sent out to the officers of the Air Force at Sourabaya: "Our sincere thanks for listening in to us. We hope you are picking up signals. All O.K."

SHOPLIFTERS BUSY

Women shoplifters were particularly prevalent in the city yesterday. From David Jones, McDowells, Farmers, Hordern Bros, and Snows, goods to a total value of over £50 were taken.

At Central Police Station, four women and one man were charged with stealing. In one case two women were charged with stealing conjointly on three counts.

DYMOCK'S BOOK ARCADE LTD., Open Every Friday until 9 p.m. Books of Every Description in Stock.

CHINESE SITE IN GEORGE ST.

TRY TO RETAIN FOOTING IN CITY.

Wealthy Chinese merchants, Gook Chew & Co., have purchased the Oddfellows' Hotel and shops in George Street, Haymarket, for £56,000.

This deal is significant in view of the fact that Sydney's Chinese have been endeavoring to hold their quarters in the city since the area commenced to brighten up twelve months ago.

According to real estate men, Chinese view with alarm the recent progress of this quarter of the city, as they fear that their exclusive area will soon be entirely occupied by Europeans.

Gook Chew and Co. did not purchase the property with the intention of running the hotel, and they accordingly sold the license to Mr. Morris for £4000. The frontage of the property is 68 feet, and the price paid, deducting the cost of the license, works out at about £788 a foot.

The license has been transferred by Mr. Morris to a new hotel, which is now being erected next to the People's Palace, in Pitt Street.

GALLANTRY OF KEITH ANDERSON

CAMERA MAN TELLS OF HIS LONE HEROISM

W. N. BERG'S STORY OF BURIAL PARTY'S EXPEDITION INTO DESERT

ALMOST CERTAIN THAT HITCHCOCK WAS FIRST TO DIE OF THIRST

PENKNIFE USED TO CLEAR RUNWAY

(See Pictorial Page).

Camera-man W. N. Berg, of "The Guardian" staff, returned yesterday from the greatest adventure—and the saddest—of his career.

A Digger, he was included in the burial party which set out into loneliest Australia to regain for civilisation the bodies of Keith Anderson and Bobby Hitchcock. That was a month ago. "It was worse than the war," he says.

Bronzed after the arduous journey in the giant Thornycroft lorry which so grandly succeeded in its job, Mr. Berg tells to-day the sad story of the quest in story and picture. His further experiences on the long desert journey will be told daily in "The Guardian."

(By W. N. BERG, "Guardian" Staff Cameraman.)

FOR a week we had pushed on through six-foot scrub of turpentine and turkey-bush. All we could see was the eternal wall of green, relieved, with the yellow turpentine blossoms, and the brownish wheels the red sand, brownish and white daisies.

We never knew from hour to hour just where we might burst the plains on us, or rounded the silent plane and its dead crew.

Suddenly the wall thinned and was gone, and we saw before us a vast desert of black ashes and red sand, on which stood row on row of charred and naked skeletons of turpentine and turkey-bush. As far as the eye could reach nothing broke the monotony of this infernal landscape.

This was at 4 p.m. on June 12. An hour later, when it seemed that the whole world must consist of fire-scarred scrubland, we burst from the green of blackened branches upon the loneliest grave a hero has ever known.

BLACKS' TERROR

The blacks, who had not known the nature of our expedition until they reached the plane, were filled with superstitious terror of the dead, and that night they slept between two guardian fires, close to our camp, instead of some distance off as usual.

Next day we rose before daybreak. Dawn was eery and almost supernatural, with a strange bluish reflection waxing over the green fringe of the unburned country, while the blackened desert around us lay still in heavy dusk.

Murray and I went to Anderson's grave, and by the time the others arrived with the lorry we had the body ready for its transfer to the coffin.

While this was carried out, the blacks were given axes, picks, and shovels, and

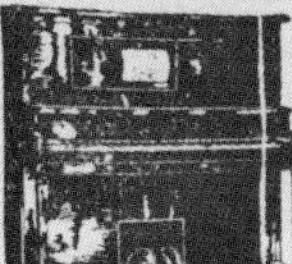

In a few moments the Thornycroft halted alongside the monoplane. She stood facing the east, tilted to the south by the depression of a punctured tyre on that side. Somehow the beauty and the grace of this man-made machine struck us poignantly as we saw her there, marooned in the pitiless waste of primeval nature.

At Constable Murray's suggestion, we all set to work at once to exhume the body of the mechanic, Bob Hitchcock, which had been buried under the right-hand wing of the plane by Lieut. Eaton's party. In the fading light I hastened to secure a few photographs to preserve a record of this grim but reverent rite.

While the party toiled at this grave, the native boys were sent out to locate that of Anderson. This was located about 400 yards south-west from the plane. Both graves were marked with rough poles cut from the nearby scrub, and the earth was heaped above them in pathetic imitation of the neat mounds of civilised burial.

By 7 p.m. Hitchcock's body had been exhumed, transferred to the coffin we had brought for it from Adelaide, and sealed within the casket.

set to work clearing a runway, as we had decided it would be impossible for us to remove the plane without severely damaging it, and, as a result, it must be flown off later.

This runway the whites improved later, and finally it extended over 300 yards in front of the plane, and was 60 feet wide, running north and south to suit the prevailing winds.

A close inspection of the plane showed it to be in excellent condition except for several vents in the fabric, caused, apparently, when the plane landed, or in the endeavors to take off again.

Extending north for about 150 yards from where the plane lay was a roughly hewn take-off through the scrub, made by the unfortunate men in their efforts to save themselves. And what a task it had been!

PENKNIFE USED!

There were absolutely no tools in the plane to assist them in this work, except a penknife, which we found in the cockpit with its blades broken and black handle scarred and battered.

Apparently the valiant castaways had made this runway with no other implements than their feet and hands.

We counted eleven separate wheel impressions running to the north, which would indicate that the plane made at least five or six attempts to leave the ground.

In the cockpit occupied by Hitchcock was found a punctured inner tube from one of the plane's tyres. Apparently this had been pierced by a spike on the ill-finished track they had made, as they started to taxi prior to rising.

The right-hand tyre was flat, and our search of the plane revealed neither a fourth tube nor a mending outfit.

This, I think, was the final blow that ended the lives of Anderson and Hitchcock. Their plane was in good condition, they had plenty of fuel, and apparently they were quite capable of controlling the machine. But a broken splinter of a scrub bush pierced their last spare tube, and cut the threads of two gallant lives.

That is my opinion, and it was that of every man there, after we had studied the evidence and considered its apparent meaning.

Why Anderson should have been found so far from the plane is a point that has not been definitely settled. In my opinion, Hitchcock, who was suffering from a badly poisoned leg, which had been dressed at Alice Springs, was the first to die, and Anderson, unable to bear the sight of the tragedy he could not stave off, wandered away from the machine, and never returned.

If my assumption is correct, it must be evidence what super-human efforts the pilot must have made to clear the runway alone after Hitchcock became too weak to help him.

A poignant and ironical sidelight is thrown on the tragedy by a revelation we heard at Alice Springs.

Several residents there, who knew the country over which he was going to travel, tried to impress upon Anderson the necessity of carrying an ample water supply, but Anderson laughed at their warnings.

Even when the last tin of petrol was being put aboard the plane, the man who was loading it pleaded to be allowed to substitute for it an equal quantity of water. But the doomed pilot lightly answered: "No thanks. Petrol is worth more to me than water."

(To be Continued To-morrow.)

ANDERSON'S BODY

ADELAIDE, Thursday.—The coffin containing the remains of Lieut. Keith Anderson left by the Melbourne express this afternoon for Sydney.

The east-west express, which leaves to-morrow morning, will take the body of Mr. R. S. Hitchcock to Perth.

SEAMAN SHOUTS HIS WEALTH

THUGS KNOCK HIM DOWN FOR £13

After being attacked by a mob, Henry Peele, 33, a cook on the steamer City of Delhi, which cleared port yesterday, looked a sorry figure in Darlinghurst police station last night.

Peele, who left his ship yesterday, set off to celebrate his shore leave with £13 in his pocket. He wandered round the vicinity of Yurong and Francis Streets, shouting and telling everyone generally that he had a considerable sum of money on his person.

At the corner of Yurong and Francis Streets, a gang of toughs set on Peele, and, knocking him to the ground, rifled his pockets.

BURST INTO TEARS

A woman companion of Peele's, seeing his plight, rushed to a taxicab depot screaming, "A man has been slashed with a razor down there."

A cabdriver rushed to the scene, and the injured man was taken to St. Vincent's Hospital, where three stitches were inserted in his upper lip.

Later, at Darlinghurst police station, Peele, when being questioned, burst into tears, and it was with difficulty that he was able to tell police the story of the assault.

For his own protection, he was detained on a charge of drunkenness.

Darlinghurst police are searching for his assailants.

RICH LAND BUT NO OWNER

REDFERN COUNCIL IN A QUANDARY

NO RATES PAID

Situated in the heart of the shopping centre at Redfern is a valuable block of land, which, according to local aldermen, has been without an owner for the last 30 years.

THE block is of triangular formation and extends about 140 feet along Botany Road and Regent Street.

It is understood that in the early days of Redfern the block of land was owned by a local solicitor, who is stated to have suddenly disappeared while in West Australia.

His son, it was understood, was the rightful owner, but from a statement made by an alderman at last night's meeting of the Redfern Council, the council has been dealing with another individual altogether in its transactions regarding the ground.

The council now claims that it has the right to sell the ground. Altogether £461 in rates is owing on the ground. This amount, aldermen state, is comprised mostly of arrears, as for the last seven years it is understood that no rates have been paid.

Before the council banned hoardings in the municipality, part of the rates were paid with the money resulting from this source.

PUBLIC DISGRACE

At present, according to a letter from the Perpetual Trustees, which was read at the meeting, the land is going to be sold privately.

"It is a public disgrace," said Ald. Holden, M.L.C., in referring to the question. "These people are only marking time with a view to securing a possessory right to the title."

Ald. J. Gilmore said that the land had been an hindrance and an eyesore to Redfern for years. "Seemingly," said Mr. Gilmore in referring to the Perpetual Trustees' letter, "Mr. Lawson Dash, whom we have negotiated with in the past, has fallen out of the picture."

DEVELOPING EMPIRE TRADE

CAMPBELLFORD (Ont.), Thursday.—Mr. R. B. Bennett, Canadian Conservative Leader, advocated inter-Empire trade which would not work to the detriment of native industry in any part of the Empire. He favored a combination of Empire trade with reasonable protection.

CHAPTER 12

The inquiry

Measured by any standards, the forced landing of the *Southern Cross* proved to be disastrously expensive. In addition to two tragic deaths, several RAAF aircraft had been lost and the public and private cost of the two searches amounted to over £30,000. In today's terms, that would be close to $1 million.

The main reason for the subsequent inquiry was not the cost, but the growing rumours that the whole affair was a publicity stunt which had mis-fired. The Government also felt that the lessons learned from the forced landings of both planes should be made a matter of public record.

The inquiry, ordered by Prime Minister Bruce at a cost of a further £2,300, commenced in Sydney on May 14th, 1929, just four weeks after the *Kookaburra* had been reported as missing.

Bruce appointed a three-man Committee of Inquiry with the authority to co-opt two more members. The committee was supported by legal counsel. Sittings were held in Sydney, Melbourne and Adelaide, with the final session in Sydney occurring on June 14th.

The chairman was Brigadier General Lachlan Wilson of Brisbane; his associates were Captain Geoffrey Hughes of Sydney and Mr Cecil Mackay, president of the Aero Club of Melbourne. The terms of reference involved the inquiry into and report on:

(i) The flight of the *Southern Cross*, its provisioning and the circumstances of the flight.

(ii) The forced landing of the *Southern Cross* and the 12-day delay in finding it.

(iii) The circumstances surrounding the *Kookaburra* and the deaths of Anderson and Hitchcock.

(iv) The loss or partial loss of three RAAF DH 9As engaged in the search.

(v) The lessons to be learned in relation to long-distance flights.

Behind this broad reference was the implicit instruction to determine if there was truth in the rumours that the forced landing of the *Southern Cross* had been pre-arranged, either as a publicity stunt or to gain public recognition for Keith Anderson.

Although such charges had been made in several newspapers, the Committee of Inquiry was more concerned about comments which had appeared in the April issue of *Aircraft* magazine. Whilst disassociating itself from the worst of the rumours, the magazine's editorial

had denounced the motives of Kingsford Smith's planned flight to England saying that it was both unnecessary and a 'newspaper stunt'.

The magazine also bitterly attacked the manner in which the media had reported the searches. In part, the editorial read:

'Nine days after the finding of the Fokker, Mr L.J. Brain, chief pilot of Qantas Ltd, in the course of his search for Anderson's and Hitchcock's machine, located it in the waterless desert, under conditions which have already been fully reported elsewhere with the most gruesome abundance of detail, piling horror upon horror by means of a faked photograph purporting to depict the body of a six-foot* man lying under a wing which (according to maker's specifications) has a folded span of only 11 feet 9 inches. To such irresponsible lengths has stunt journalism descended that the "body" is shown occupying only one eighth the length of the wing.'

The same magazine also implied that Ulm had managed to get his diary to the outside world and had arranged for a writ to be issued on his behalf at a time when he and Kingsford Smith were supposedly unable to communicate with the outside world. The magazine further claimed that Smithy and his crew had been within walking distance (40 kilometres) of Port George Mission, but had elected to remain where they had been forced to land for two weeks.

These, and the original charge that Smithy and Anderson had colluded in a 'lost and found' operation, created immense public interest. The Sydney *Sun* proudly announced record sales of 335,000 compared with a previous high of 284,743 which had been attained a year before during a fiery state election.

One allegation — that Anderson was flying with a sealed order (presumably the location of the *Southern Cross*) — was defused before the inquiry started when Anderson's backer, John Cantor, made a public statement. He said:

'From the very moment that the *Southern Cross* was reported missing, I felt sure — but I don't know why — that the plane would be found where, as a matter of fact, it has been.

'My reason for not divulging the order I gave Lieutenant Anderson was that I did not want anyone else to get there first. I wanted Lieutenant Anderson to find the *Southern Cross*.

* *1.83 metres*

'That is all there was in my sealed order. No one knew what range the *Kookaburra* had, but I can tell you that she should have made her hops without any trouble.'

Other statements came to light. Bon Hilliard had expressed doubts that the aviators were really dead, when they had been found, so she sent an urgent telegram to Sir William Glasgow, secretary of the Air Board, saying: 'Thanks for wire but seriously alarmed at conclusion fliers dead without examination medical experts consulted think otherwise.'

However none of these matters affected the basic purpose of the inquiry which was to find out how and why the *Southern Cross* had become lost in the first place.

During the sessions, Kingsford Smith vehemently denied what the *Sydney Morning Herald* had called the 'base rumours'. He said the story that the forced landing had been a premeditated publicity stunt was an 'absolute, deliberate and malicious lie'. When asked about a published statement that there had been an arrangement under which the fliers would lose themselves and then be found by Keith Anderson, Kingsford Smith angrily shouted: 'That is another deliberate malicious lie which affects a dead man's reputation and which I consider disgusting. It is disgraceful.'

In the course of evidence, he said that the alleged 'mysterious message' which had been handed to him at Richmond was merely confirmation of a telegram which had been sent the previous day from Wyndham relating to 'fine weather' and a 'drying wind'. He also claimed that several attempts had been made to radio their approximate position from the Glenelg flats but, as the plane was not in the air, they had little power to drive the generator, so the signals had not been received. He added that the bright silver plane should not have been difficult to spot from the air as it had a wingspan of 72 feet (21.6 metres). However the search planes had flown over them twice, despite their smoke signals.

There was little humour evident during the inquiry, but one of Smithy's answers broke the tension. Testifying that he had risen from bed after a bout of influenza three or four days before the ill-fated flight, he said that his doctor had declared him to be fit to start. 'Did you get a certificate to this effect?' he was asked. 'I could get one. All I have at present is a bill,' he replied.

Concerning an allegation that Kingsford Smith had endorsed a promissory note for £300, (this would be worth about $9,000 in terms of today's buying power), as part of the agreement, Kingsford Smith told the inquiry: 'That is a malicious lie. I endorsed the promissory note purely as a matter of friendship.'

Smithy proved to be a spirited and convincing witness for his own defence, but some damaging claims and counter-claims surfaced. The first exploded when William A. Todd, a former officer on the *SS Tahiti*, testified. He was on the ship which had carried Kingsford Smith, Charles Ulm and Keith Anderson to the USA before they had purchased the *Southern Cross*. Todd claimed that after the journey he had met Charles Ulm on several occasions in San Francisco. Once, over drinks, Ulm had said that if he and Smithy had become lost during their around-Australia flight, they would have created such a 'whale of publicity' that they would have secured all the money needed to sponsor further flights.

Todd's reliability as a witness took a battering when he admitted to being a heavy drinker and that he had wrecked a new car on the day on which he claimed that Ulm had made the remark. Todd also admitted that Smithy had asked him to navigate the *Southern Cross* on the epic Pacific flight. Todd had agreed at first then later declined the offer, since he believed that the flight would never eventuate. Further, he admitted that he was disappointed when he had not shared in the eventually successful flight. Todd's evidence was supported in part by the manager of an Australian oil company, Dudley Lush, who told the inquiry that Ulm had spoken to him some months earlier, suggesting that he and Smithy should 'whip up some publicity' by becoming lost in the desert. Lush admitted that he had never heard a similar suggestion from Kingsford Smith.

Formal evidence was given on the exchange of telegrams and about a confusion over deliveries, which prevented Kingsford Smith from receiving the true picture of the prevailing weather conditions. Independent witnesses testified that no experienced flier would have deliberately attempted to land a valuable plane in the mud-flats. The fact that Smithy had brought the plane down intact was described as a 'wonderful achievement'.

Other evidence showed that if the affair was a publicity stunt, it had been very badly planned. The near empty food lockers in both the *Southern Cross* and *Kookaburra*, certainly did not indicate that the men expected to be stranded in the desert.

Had Smithy arranged to be 'found' by Keith Anderson on the banks of the Glenelg River, it would have been a near miracle for him to fly through atrocious weather, become lost twice, navigate the predetermined route and finally descend at an appointed place with the fuel tanks almost empty. All this was supposedly accomplished without a radio to aid them.

Since Kingsford Smith and Ulm owned the *Southern Cross*, then worth about £10,000 (about \$300,000 in today's terms), another convincing argument was presented. Only a fool would have risked damaging this investment in the vague hope that publicity would generate increased sponsorship. Kingsford Smith was anything but a fool and he already had a sponsor, Atlantic Union Oil, for the trip to England.

A little-known argument is that the Governor-General, Lord Stonehaven, had told Smithy a few days before the flight in question, that he would be presented with the Air Force Cross (which had been awarded to him earlier), upon his arrival in England. The King would be making the presentation if he had recovered from an illness, otherwise the Prince of Wales would take the role. Everyone who knew Smithy and the immense pride he took in his achievements, realized that he would not risk losing such an honour, to participate in a cheap and dangerous publicity stunt.

Other evidence made it clear that Anderson was probably motivated by a genuine desire to save Smithy's life. Anderson himself had been lost twice in Western Australia — once for three days and once for three weeks — and on each occasion, Smithy had found him.

Captain L.W. Sutherland of Vacuum Oil Company stated that he had discussed the flight with Anderson before he left and Anderson had marked a spot on a map where he said the missing plane would be found. The spot was 160 kilometres from the actual Coffee Royal Affair's landing site. However it was also shown that many of the radio signals despatched from the *Southern Cross,* when it was still in the air, had been picked up and Anderson, with his extensive knowledge of flying in Western Australia, had no trouble in plotting the plane's route and forming his own conclusions as to its location. Very likely his conviction that he could find the *Southern Cross* had encouraged him to 'throw caution to the winds' and proceed hastily and unwisely.

The inquiry produced some interesting revelations.

Colonel Brinsmead, controller of Civil Aviation, told Mr Hammond (counsel for the committee) that his branch would have done everything possible within its power to prevent Anderson from making his flight across the desert, because the machine would necessarily have been overladen when carrying so much petrol. When pressed on the point, he said that had he known Anderson's route, he would have wired the local police authority to say that unless the Department received Anderson's assurance that he would only fly via Newcastle Waters and Wave Hill, his pilot's licence would be suspended and, if necessary, the registration of his machine withdrawn.

When Brinsmead was asked why, he answered:
'Because it was quite unnecessary to fly in that way to get to that point. The way it could have been done, in which it was proposed to be done afterwards by Brain, was to fly north from Alice Springs to Newcastle Waters, and then turn west at Newcastle Waters and fly in the direction of Wave Hill. That would have meant about an additional one and a half hours flying ... and it would have been easily possible for him to come down at an intermediate place to refuel in case of necessity.

If we had known he was flying north and then west to Wave Hill, we could have sent a machine along there and we could have picked him up with practical certainty, but flying in the direction he did from Woodford Crossing, across 550 miles* of desert country, if he had come down in some part there, it was almost impossible to pick him up.'

Eaton in his evidence, estimated that the burned out bush extended over about 100 square miles (25,900 hectares). He described the desert as 'covered with low brush, thick in some places, thinning out in others, with ant hills and occasional odd trees about 30 or 40 feet** high in places'. He added that the soil was sandy and that there was no water, so far as he could see for 45 miles (72 kilometres). He said that there were no animals in the area, adding:
'There was one rather gruesome test of that — the bodies were not touched in any way at all.'

In his evidence, Eric Douglas stated that after thoroughly examining the *Kookaburra,* he had tried to start the engine and it had fired up on the 'second pull'. It was running evenly on all cylinders, and he examined carefully the faulty lock-nut. He found it had been correctly adjusted and the spanner had been left on the ground under the engine.

Douglas said that the *Kookaburra's* instruments consisted of a revolution indicator, air speed indicator, an altimeter and wind indicator, and an oil pressure gauge. He said that the holder of the compass was lying in the rear cockpit with the bowl. The compass card had disappeared and part of the compass frame was on the ground. He said that he could not test the unit as it had been completely wrecked.

Douglas gave evidence about the *Kookaburra's* wheel marks, saying that there had been two or three attempts at take-off. Apart from damage to the fuselage, he said, the machine was quite airworthy.

On the question of tools, he disagreed with other statements that the *Kookaburra* had no tools, stating that there was a tool roll in the

* *880 kilometres.*
** *Nine to 12 metres.*

One of the points which the Committee of Inquiry discussed, concerned the reasons why the *Southern Cross* was not found for 12 days after it had been forced to land. The *Southern Cross* was the indirect cause of Anderson's and Hitchcock's death.

front cockpit consisting of a propeller spanner, box spanner, three or four set spanners, pliers, shifting spanner, bicycle tyre pump and a spare tube. He said that no attempt had been made to fit the spare tube.

Douglas made the point that the 'runway' had been fairly well cleared for 450 feet (137 metres) and the bush had been broken down for a further 150 feet (45 metres). He said that the wheel tracks went along this 'runway' for about 420 feet (128 metres) and he reasoned that the plane did not actually leave the ground because there were no marks to show that it had come down again. Further he pointed out that the aviators had cleared the runway with their hands and boots.

Martin Kromer, a prospector, said that he was at Algebuckina on April 9th, where he saw a plane which flew past Algebuckina but turned back to land. He went to see what the problem was, and then helped the aviators to 'do up' the engine. He mentioned two men, one was called Anderson, the other Bob. He said that when they started the engine, two or three lumps of cotton waste came out of the exhaust pipe.

'Bob reckoned that was enough to cause the trouble,' he told the inquiry. Since there were few tools, he said, Bob was forced to undo a lot of nuts without spanners. This may have been incorrect since spanners were found in the *Kookaburra*.

'He had to do the work with a screwdriver, and he had something like a corkscrew to knock them off with.

'Mr Anderson said, "I thought you had a good set of tools", to which Bob replied, "Someone stole them while I was in hospital". Anderson said, "Why did you not lock them up in your locker?", to which Bob replied, "I did but while I was in hospital someone wrenched the lock off and stole the tools".'

Mr Laurie Norman Phipps, a motor mechanic and licensed ground engineer, was called to give evidence. He stated that he had flown to Richmond with Anderson on April 6th. When asked if tests had been made on the compass, he replied:

'Keith Anderson went into the orderly room and examined a chart with Squadron Leader Lucas and discovered which way the joints of the tarmac ran ... the machine was placed running along those lines which was a north point. The compass was then adjusted to the north point.'

In answer to questions, he said that the plane was brought around to the east point, once again using the chart in the orderly room, but there was not enough time for full testing to be carried out.

He said that Keith Anderson was 'quite satisfied with the compass', and the method by which it had been adjusted, but no deviation chart had been prepared as it was just becoming dark. Phipps added that tests had been made with about 40 gallons (182 litres) of fuel in the tanks. Anderson had taken off and circled the aerodrome for about 12 minutes. As he had made two flights, he then pronounced himself to be satisfied with the plane.

When asked what tools were in the plane on April 6th (the day before take-off), Phipps said that he had prepared the tool roll, which was a standard kit and in addition, 'I supplied him with my own personal spanners. He had one adjustable spanner (nine inches* long), one plug spanner, one oil filter spanner, one coil drain spanner, one small King Dick, four set spanners, one punch, one screwdriver, one hammer, one grease gun and one magneto spanner.'

In answer to further questions, Phipps agreed that there had been no other tools on board. With regard to supplies, he testified that the plane had taken off from Richmond with two pints (1.2 litres) of water, in clear distilled-water bottles. Further, he had personally tied up two packages of sandwiches for the men.

The committee returned to the question of the faulty compass. When asked if the nearby galvanized iron shed would have affected the correct swinging of the compass, Phipps replied that it would have little effect as 'Keith Anderson's machine had 16 pounds** of steel in his tool roll and spare parts and it took four needles to true the compass out'.

Asked why the compass was not registering the correct directions after Broken Hill, he replied:
'The reason it was so much out was that his tool kit was stolen and also his spare parts.'

'You say that would affect the compass?' asked Mr Hammond for the Committee of Inquiry.

'Yes. The reason that stuff was put in the plane was to counteract the effect shown by the needles.'

On the subject of fuel tanks, Phipps said that he had fitted an extra tank at the front to carry 39½ gallons (180 litres) and one at the rear carrying 29½ gallons (134 litres), five weeks before the plane left for the search. He said that this would have given the plane a 20-hour

cruising range but no tests had been made with a full fuel load. Asked what was the maximum permissible load for the machine, Phipps replied that the amount was 1,600 pounds (727 kilograms). He agreed that the plane's weight, when fully laden with fuel for a flight from Alice Springs had been 2,080 pounds (945 kilograms).

Mr Hammond pressed the point, by asking, 'Do you consider as an engineer that it was an airworthy plane?' The answer was, 'It was and it was not. If we were to stick to the red tape of the laws of aviation it was not, but the fact that he was able to take off proves that the plane was airworthy.'

Phipps' other answers indicated that Anderson had flown the *Kookaburra* for only two and a half hours prior to leaving Richmond and that it had been flown only once since the additional petrol tanks had been fitted. He also made this revealing statement:
'It was not prepared for this flight at all. She was unpacked on the 23rd of December. This was done for special reasons of Keith Anderson's own. The machine was bought for him for other reasons than the flight. On the 24th of December, after she had been assembled, he flew the machine for two and a half hours. On the 4th of March, he gave me instructions to go right ahead with the fitting of the extra tanks which was to be done for the purpose of endurance records. I kept it for five weeks fitting it up. On Friday the 5th of April, Keith Anderson rang me and said he would be getting away on a search on Sunday morning if possible ... the compass was brought out at about 2.15 and I put it in the machine. The compass had been fitted at my place. We took off immediately after the fitting in of the compass, that is, rivetting it in etc. We took about quarter of an hour to do this and that meant we took off about 2.30 and we arrived at Richmond at 2.55.'

Phipps also said that Hitchcock had not previously flown in the *Kookaburra* or a similar plane and that he appeared very weak. He was 'hardly able to lift a tin of benzine to the centre section for filling the tank', said Phipps.

As the questioning developed, Phipps became increasingly confused, at times contradicting his own evidence. It is clear however that the bundle of tools, placed close to the compass, had a significant effect on its accuracy and that, very possibly, the movement of these tools around the cabin may have swung the needle by an appreciable amount. After questioning, Phipps admitted that the compass had been incorrectly swung and that he shared at least some of the blame. He also agreed that when Anderson took off from Alice Springs with a full load, it was the first time that the *Kookaburra* had carried such a weight.

---

23 centimetres.

**7.3 kilograms.*

Milton Kent, who had sold the *Kookaburra* to Anderson, testified that the *Kookaburra's* fuselage and undercarriage had been specially strengthened. The new axle was about twice the weight of the standard axle and the machine was also equipped with 'double cross-bracing cables'.

John Cantor revealed that a Mr Douglas Kerr had approached him with Keith Anderson, soon after Smithy's plane was reported missing, and said:
'Jack, Keith is dead anxious to go away and look for Smithy.' After some discussion about the *Kookaburra,* Anderson said that he needed a sum of £500 to enter the search. (Evidence later showed that this sum was to cover all expenses, including a fee of £100 for Hitchcock's services.)

'He left me then,' said Cantor, 'and a little later I approached a number of the boys, friends of Smithy.' About 12 men agreed to back Anderson's search and they left it to Cantor to tell Anderson that the money was guaranteed and he could prepare for an early departure. Anderson is quoted by Cantor as saying:
'Jack, I think they are looking in the wrong direction. I have an idea that they are down around Kimberley or Port George. The planes are looking up in the Drysdale area but of course they must be in the other direction.'

A sworn declaration was received from the Marree postmaster stating that he had handed Anderson a telegram received from Colonel Brinsmead and that after reading it, Anderson had said aloud:
'This has nothing to do with my search. I am flying under private arrangements and intend going on.'

Edward J. Hart, editor and publisher of *Aircraft* magazine, was called and he took a lacing from the counsel acting for the committee. Under questioning, it became clear that many of the damning statements he had published were opinions of his own, which were not necessarily supported by fact. It was also revealed in evidence that he had had several arguments with Ulm in the past and that his motives in publishing the attack in his magazine were therefore suspect. His evidence did not appear to impress the committee in the least.

•• •• ••

The committee's conclusions

The Air Inquiry Committee began its public sittings in Sydney on May 14th, and finished on June 14th, 1929. The report, published at length, exonerated Smithy and Ulm and stated emphatically that there was no evidence to suggest that the forced landing of the *Southern Cross* had been premeditated and there was nothing to impugn the honesty of anyone aboard. The committee called Smithy's decision to fly on after losing the aerial an 'error of judgement' and heavily criticized him and radio operator McWilliams, for failing to make adjustments to the radio receiving set so that it could have been used as a transmitter. The chairman, Brigadier General Wilson, also criticized the crew for not checking the emergency rations before leaving Richmond.

He stated that the *Southern Cross* crew should have used the 18 gallons (82 litres) of spare engine oil which they were carrying to supplement the fires and create dense smoke to assist the searchers and 'their failure to make use of this valuable medium is inexplicable'.

The inquiry found that the *Kookaburra* was ill prepared, with no radio, emergency rations or adequate water supply. It noted that the *Kookaburra* was heavily overladen when leaving Alice Springs and although it carried tools for effecting machine adjustments, there was no tomahawk or similar implement.

It also found that the *Kookaburra* was a machine of a type which was not suitable for the proposed flight and that its compass had not been properly swung and was unreliable. Anderson had not been willing to allow the necessary time to enable it to be correctly adjusted.

The committee considered that the compass was the primary cause of the tragedy but that the *Kookaburra* could have followed the overland telegraph line. Had it done so, it would have 'thus been in touch with that line and able to tap in if the required instrument had been carried but Lieutenant Anderson decided to save about one and a half hours of extra flying by going across the desert'.

The findings went on:
'In view of the previous experience of both of these aviators in long-distance flights, the committee can only come to the conclusion that Lieutenant Anderson was so anxious to start on the search for his old comrades and friends and so anxious to carry every possible gallon of petrol, that he would not wait to have his compass properly tested and would not carry an extra pound of rations or tools, deliberately taking the risk of any accident.'

The committee highly praised 'persons who with single-engined machines were engaged over long periods in searching for the *Southern Cross,* under circumstances which were extremely dangerous to the searchers in the event of any engine trouble to their machines and in this regard mention Messrs Woods, Heath, Chater, O'Dea, Holden, Siddons, Chateau, Davidson, Dr Hamilton, Stannage and Mitchell'.

Smith's Weekly

PRICE 4d

FOUNDED BY JOYNTON SMITH — The Public Guardian

Net Sales Per Issue Exceed 200,000.
Sydney "Guardian" audited net sales 169,613 daily.

Vol. XI, No. 12 (Copyright)

Saturday, May 4, 1929

Registered at the General Post Office, Sydney for Transmission by Post as a Newspaper.

"SMITH'S" VAUDEVILLIANS HAVE A FEW WORDS TO SAY ON ACTING

Barrage of Questions Inquiry Committee will Put Down!

ANSWERS MUST BE FOUND

The "Southern Cross" and the Aftermath

POINTS TO BE LINKED UP

"SMITH'S" gives below, with greater accuracy of detail than in any other newspaper in the Commonwealth, the questions to be asked by the Inquiry Board, concerning the "Southern Cross" fiasco and its disastrous consequences.

Important comparisons will be made on the question of equipment, and on the relative efficacy of the search for the "Southern Cross," with a living crew, and for the "Kookaburra," whose pilot and mechanic were dead.

OFFICIAL *"terms of reference" to the Board of Inquiry include discussion of the following:—*

(1) Adequacy or inadequacy of "Southern Cross" provision of equipment, accessories, and food;

(2) Reasons why the "Southern Cross," after landing, was not discovered between March 31 and April 12;

(3) Loss of the "Kookaburra," with reported deaths of Anderson and Hitchcock;

(4) Loss of the Commonwealth aeroplane DH9A;

(5) Desirability of better precautions for long-distance flights.

Kingsford Smith in the desert.

Ulm taken at the same time.

SUCH is the skeleton outline of the investigation the Board will undertake.

The Commonwealth Government, however, will indicate to the Board, in more searching and direct terms, some of the precise questions to which answers must be found.

(1) Why did not Smith and Ulm's party of four, the morning after that forced landing, make efforts to climb the encircling mountains, in order to ascertain their position and their bearings.

At that time they would still be hale and hearty, and perfectly able to accomplish the feat. Complete exhaustion fell upon them quickly; for Ulm reports in his diary of April 2 (the third day) only:

"Broome sent us a message by light a triangle of three 100 yards apart; this will be impossible in our present weakened condition."

(2) As Ulm stated in his diary that they had a certain amount of petrol, why did they not endeavor to generate power, so as to get into touch with the outside world by transmitted radio?

According to Ulm's diary of April 2 (the second day) there was at least enough petrol to let the machine stay in the air "for fifteen minutes."

Experts will be asked to ascertain, for how long a period of time this admitted supply of petrol would have allowed transmitting radio to be generated.

In the view of the Commonwealth Government, it is patent that if such radio had been sent out, the "Southern Cross" on the Glenelg flat would have been located almost immediately, and the crew succored, without the loss of £25,000 and two lives.

(3) Why did not Smith and Ulm, being in receipt of wireless messages, try to create smoke signals by using oil and petrol, which in the main are excellent ingredients for that purpose?

More than once in his diary Ulm complains of the difficulty or impossibility of giving signals.

On Saturday, April 6, a plane passed "not more than five miles away" but though all four men on the ground "worked furiously, stoking up a big smoke fire," they failed to attract attention.

On April 9 a plane passed twice, at an interval of seventy minutes (9.30 a.m. and 10.40 a.m.), from 3500 to 4000 feet up, "on a clear cloudless day."

"We had lit a large fire near the mangroves to the south," says Ulm's diary.

In this connection Pilot Brain in the "Atalanta" in the Northern Territory saw smoke from 6000ft at a distance of 60 miles, and from 2000/2500 feet sighted the "Kookaburra" on the ground at seven miles.

(4) Why did not West Australian Airways act in accordance with the suggestion of Colonel Brinsmead, and make their base at Port George instead of Wyndham?

This question appears to have been raised to cover the attitude of the Defence Department in their actions towards the affair.

When it was asked as to why the Department had not taken action sooner, the answer was tendered that there were no suitable machines available.

It is the policy of the Air Force for machines to work in pairs, but the range of the machines available was not sufficient to allow of two machines to work in consort.

(5) Was the necessary attention given to the equipment to be carried for the flight from Australia to England? If so, why was it necessary for Smith and Ulm to use screwdrivers to cut down a tree, and, further, why should Smith state that one screwdriver was used and Ulm say two?

Further points that the public would like to be informed upon, and that will doubtless be raised, are these:—

(a) Was Kingsford Smith, as a former pilot of the Western Airway, presumed to know the coast and the conditions thoroughly?

(b) What quantity of petrol did the "Southern Cross" carry on leaving Sydney, and what quantity was discovered to be on board after replenishing supplies at Wyndham?

Was there not, in any case, sufficient petrol in hand when the plane landed on the mud flat to allow of a circling flight to ascertain their position?

Might not this position have been expected to be ascertainable in view of Brain's statement that visibility from his plane extended to a radius of 70 miles, even allowing for differences of visibility, due to coastal conditions?

What equipment in the ways of tools did the "Southern Cross" carry on the Pacific flight?

Regarding Flight-Lieutenant Eaton, who is in charge of the party which reached the Kookaburra last Monday after a seven days' march to seek the bodies of Anderson and Hitchcock, it is learned from official sources that he will probably be recommended for a decoration. The fact has not been published that, when he crashed a few days ago, he walked nine miles and put through his official reports, some of which were lodged after midnight.

He suggested in one of these reports that he fly over the spot again and land by parachute.

This was given careful consideration by the Department, and he was advised to consider the risk. Whatever he and Sergeant Douglas have done has been on their own initiative, and their actions are thought very highly of at headquarters.

(See page 3 for more Air matter).

Prehistoric fire-water bibber: Eh, dash phizzuph out of order again!

BRAINS will out.

THERE is an ape in the Zoo with a 10-foot arm stretch. And he doesn't know even one fish yarn.

SHEARING has started in the Tibooburra district. It finished with the Show in Sydney district.

OBJECTING to the smell of a horse's breath, 400 men laid N.S.W. colliery idle. Breathless argument.

COAL Crisis. —100% talkie.

MANLY'S "sand modeller" is going to America to try his luck. There's grit for you.

"GRASSHOPPERS" is said to be the latest color for dresses. Good for jumpers.

SCOTSMAN claims to have worn the same kilt for 30 years. Off and on, of course.

MARCHIN' PLACE.

CATCHING criminals.—Child's play.

THE Cold Snap.—Kodaking on Kosciusko.

MELBOURNE "Sun" serial: "My visitor pulled out his pipe and filled it with savage fingers." Put that in your pipe and smoke it!

OF nights our flappers get all dressed up and know where to go.

Mrs. Naggs: Do you have to put any water with the cough medicine your husband is taking?
Mrs. Maggs: No, onions.

"SUN": Man, dead two years, served with summons.—Stiff!

CHORUS girls enjoy their work.—Get a kick out of it.

TO get a meal, a man threw a stone through the window of a police station. Bread for a stone.

ACTRESS says bed scenes are necessary to sustain interest in modern plays. Bunk.

FRANCE has cancelled the vaccination order!—No 'arm done.

SYDNEY High School.—He ROWS.

WANTED: Plane Facts.

THE talkies have come to stay.

GENERAL BLAMEY (Vic.), says the police horses are being replaced by motor cycles. Couldn't stand the chaff.

KNIGHTHOOD for Jack Hobbs? Best reward for centuries!

NAVVIES at railway camp drink "Flytox."—News item. "One with the flies."

A MURDERCA.

VAGRANT in Melbourne admitted he hadn't worked for ten years—a Mel-bum.

"SHOT in the night."—"Daily Guardian." Sly grog?

A KNIGHT in Toone.

SIR Benjamin Fuller said his theatres were not afraid of the talkies. Fuller confidence.

AFTER 50 years' occupancy, the "Edina" has been ordered to vacate her berth on the Yarra. Eener, deener, 'dina, go.

GRETNA GREEN couple were married by the village blacksmith. Under the wedding chestnut tree?

J. B. SOBBS.

"TIMBER worker assaulted with length of wood by strikers."—News Item.—Via Beam.

Lady Motorist: Haven't I met you before somewhere?
Victim: Very likely; your bumper bars seem familiar to me.

"THERE are boys licking stamps in Australia today," says Sir Arthur Duckham, "who will be millionaires some day." Not if the stamp account is audited properly.

ENGLISH telephone girl becomes a nun. Changed her "hello" for a halo.

JOHN JALE, of Queanbeyan (N.S.W.) has just celebrated his 99th birthday. Gale and hearty.

J.F.O.: Gibson's Poor Office.

STATE Bakery at Stanmore (N.S.W.) has been sold for £9000. A decent roll.

On the subject of the *Kookaburra* search members of the committee 'particularly wish to commend the splendid efforts of the various Air Force pilots who were dispatched by the Federal Government to find the *Kookaburra*, also Mr L.J. Brain of Qantas, who, with his crew (Mr Stephens and Mr Compston) subsequently found the *Kookaburra* south-east of Wave Hill'.

Smithy and Ulm emerged from the inquiry exonerated but with their public image tarnished.

Anderson and Hitchcock, by contrast, became national heroes — men who had died for their mates.

Aftermath

The various men involved in the search for the *Kookaburra* were also honoured officially and unofficially.

Lester Brain was involved in another widely publicized search two months later. Aviators, Moir and Owen had set out to fly from England to Australia in a single-engined Vickers Vellore. They had reached the Australian coastline and were forced to make a crash landing near Cape Don lighthouse. Although they were in the care of the lighthouse keeper, he had no way of contacting other people. Brain found the missing plane and the men were duly rescued. Immediately afterwards Lester Brain received the Air Force Cross.

In his book *Qantas Rising,* Sir Hudson Fysh says that he reported to his board:
'Owing to the nature of the trip, the publicity which we received amounted to a tremendous boost for Mr Brain and Qantas. The helpful publicity which we have received through the success of the two air search trips could probably not have been bought for any money.'

After World War II Lester Brain was appointed as first general manager of TAA and later managing director of the De Havilland Aircraft Company in Australia.

The RAAF pilots were also highly commended by their departments.

Subsequently, Ulm and Smithy both died when attacking long-distance flying records. Ulm was flying from California to Australia when his plane vanished at sea. He was never seen again. Smithy — after an incredibly successful record-breaking career — was even-

tually lost when attempting to break the England-Australia air record in his Lockheed Altair. Only one wheel and an undercarriage strut from his plane were found.

At the time of the Coffee Royal incident, Smithy and Ulm were establishing their own airline, Australian National Airways.

Despite an ordeal which, including the inquiry, lasted for two and a half months, Smithy was determined to seek greater glory. Within weeks of the close of the inquiry, he and Ulm flew the *Southern Cross* to England in the record time of 12 days and 18 hours, arriving on July 8th, 1929.

They battled for many years seeking air mail contracts, but none was awarded. The airline went into liquidation in 1932, after one of its planes — *Southern Cloud* — crashed in a storm, killing six passengers and the crew of two.

For the rest of his life, Smithy tried to secure air mail contracts, and no one knows why he was so totally unsuccessful. Strangely enough, one possible reason has received little notice. The political activities of both Charles Kingsford Smith and Charles Ulm were disturbing the Government of the time. Both were members of the New Guard, a right-wing organization founded in 1931 by Colonel Eric Campbell. In many respects it followed the para-military thinking of Mussolini's Fascist group and Oswald Moseley's Fascisti of Britain. Colonel Campbell, commander-in-chief of the New Guard, and his New Guard colleague, Sir Frederick Stewart were both directors of Australian National Airways.

Although the New Guard was not a militant group, it was distrusted by the UAP Government of Joseph Lyons and the Federal Labour Government of 1929 to 1931. This distrust may well have been the reason why Australian National Airways met official hostility to their applications for Government air mail contracts or subsidies. Qantas already operated with a subsidy, and was highly regarded in Government circles.

Bon Hilliard never married her fiancé, and in 1935 she married Major Thomas Tate. He died 11 years later, but Bon never remarried. Anderson's mother featured in the news at the time of the tragedy. Her local member of parliament complained that she had not been receiving regular news during the search for her son and had been left to rely on newspaper reports. A cable was sent on her behalf to the Defence Department and arrangements were made for direct news to be sent to her and Bob's estranged wife, Violet.

Violet Hitchcock lodged several complaints. Even before Eaton's ground party had reached the scene of the forced landing of the *Kookaburra,* she had written to the Prime

Minister, Mr Bruce, saying that she had heard that her pension would be cancelled.

One of Bob's sisters cabled the Defence Department saying:

'Mechanic Hitchcock family living at 46 Money Street, Perth, would appreciate any definite news when received. Have reason to believe blacks have got him. Signed Mrs Davis, sister.'

The prospect of receiving news through Mrs Hitchcock senior did not please Violet so she wrote to the secretary of the Defence Department saying:

'As some confusion seems to exist about my address, I am writing to ask you to rectify the error. I am the widow of mechanic Hitchcock and the telegram telling me about his death was sent to 46 Money Street, with the result that the world knew it before I did. I am not, or never have been, in residence at 46 Money Street, so I would consider it a favour when communicating with me if your department would address me as above (592 Albany Street, Victoria Park, WA). Thanking you in anticipation, I remain, Yours sincerely, Mrs Violet E. Hitchcock.'

She followed this with the message that she wished her husband to be buried in Perth.

Mrs Violet Hitchcock also made a formal request that the effects of her late husband should be forwarded to her. Mr Eaton had written to inform her that Bob's gold watch and chain were not found on his body. There was consternation when the director of the Investigation Branch, acting under orders from the Prime Minister, disclosed that one of his officers had located the watch and chain, with a small gold nugget pendant at the pawn shop of a Mr P. Davis of Paddington. However an investigation showed they had been pawned the previous March before the fatal journey for the sum of £2/10/0.

The Government redeemed the articles for £3/2/6, including interest. It was also disclosed that Mr Hitchcock had launched a small debts case against a Mr A. English of Goulburn for £16/18/0 for payment of work done, but the case was struck out when neither of the parties appeared in court. The Investigation Branch made enquiries at the Customs House Hotel and learned that John Cantor had paid a Mrs Baker the sum of £100 on behalf of Mr Hitchcock. This money was paid after the fatal flight commenced and was due to be repaid to Cantor from Hitchcock's earnings during the flight.

Bob's brothers, sisters and parents were heartbroken by the news. His sister Pretoria's husband and her brother-in-law were wharf labourers at Fremantle in Western Australia, when news first came that Bob was missing in the *Kookaburra*. They went to their union's secretary and explained the position. He contacted John Curtin, then the federal member for Fremantle, and it was Curtain who eventually arranged for the body to be brought back to Perth.

At one stage this was certainly in doubt, and Stanley Hitchcock wrote to the Prime Minister on behalf of his parents, brothers and sisters, saying that he hoped that the Prime Minister 'will keep your promise and have our hero brother and son's body brought back to Perth as soon as circumstances permit so that we will be able to pay our dear one last respects. We all feel our great loss very keenly...'

At the time of writing this book, Les Hitchcock, one of Bob's sons, is still alive. He was 15 years old at the time of his father's death. Bob, the other son, was killed in an aerial accident in August, 1940. Piloting a Lockheed Hudson converted bomber, he was bringing seven senior Government ministers and two service chiefs to Canberra for a major conference. According to reports of the day, 'the pilot of the plane inexplicably dipped its wing to port and crashed into a hill close to the airport, killing the passengers and the crew of four'.

Two other parliamentarians, assistant Treasurer, Arthur Fadden and Senator McLeay had made last-minute withdrawals from the passenger list. The Minister for the Army, Mr Street, (father of Mr Tony Street, the Industrial Relations Minister of the Fraser Government) was one of the victims.

Bon Hilliard wrote a sweet note to the Prime Minister saying:

'I wish to thank you sincerely for all you have done and are doing for my dear Keith and now I want to tell you that it is my earnest wish (and I know it would be his also) that his plane the *Kookaburra*, his cherished possession, be placed in the National Museum and there given great care.

'Also I would like Keith's writings on the rudder of the plane to be protected — may I suggest that the writing be covered with mica if possible. His writing would still be legible and yet protected from interference in any way. I know you will understand just what I wish and possibly you may think of some better idea, so I shall leave it entirely in your hands. Thanking you Mr Bruce, Yours very sincerely, Lyal Hilliard.'

Richard Jackson of Blayney wrote to the Citizens Committee saying that he had loaned Bob Hitchcock a leather flying coat, which, for sentimental reasons, he greatly treasured. 'The subject of this letter is a very painful one,' he wrote, '... but if there is any way in which I am able to eventually regain possession of my coat, your assistance will be appreciated.'

Twenty-six pounds had been found close to Anderson's body and this was paid into his estate. The Air Board also made an inventory of the personal effects of both men, including the 'diary', the plane's joy stick, assorted clothing including the leather jacket, a bible, various maps, an unsigned copy of the agreement between Anderson and Cantor, a pair of spectacles, an open order authority with Vacuum Oil, the receipt for the *Kookaburra* (on the back of one of Milton Kent's business cards) and several telegrams. The packages were distributed to Mr A.V. Hilliard, Mrs Anderson and Mrs Hitchcock.

The aircraft was virtually Keith Anderson's only asset. Mr Hilliard stated that failing action by the Government to recover it, he intended to take the necessary steps himself. Mrs Anderson did not publicly state her views but Bon Hilliard let it be known that she wished the plane to go to the War Museum in Canberra. Several offers were made to buy it, including one from Mr Charles Berg of Newcastle Waters.

When asked by the Government for his advice on the question of recovery, Eaton stated that the only practical method of salvaging it would be to approach it by road. By that time it had, he said, been standing in the sun for some months and the wooden parts would be partly warped. This meant that a thorough inspection would be necessary before an attempt could be made to fly it out. The secretary of the Department of Defence advised that the recovery cost by road would be at least three shillings per ton per mile and it was doubtful if a suitable trailer would be available at any place nearer than Adelaide. He said that the cost would be out of all proportion to the value of the plane.

The secretary of the Defence Department finally advised the Prime Minister as follows: 'In the circumstances, it is recommended that A.V. Hilliard, 84 Pitt Street, Sydney, who is acting on behalf of Mrs Anderson, mother of the late Keith Anderson, be advised that opinions expressed by Flight Lieutenant Eaton and the Thornycroft truck party are to the effect that the salvaging of the machine presents almost insurmountable difficulties, apart from the considerable expenditure which would be necessary should any attempt be made, and that the Government therefore, while fully appreciating the historic and sentimental value of the machine, regrets that it cannot see its way, in the present state of the Commonwealth finances, to allocate any funds for the purpose of the recovery of the plane.'

Postcript

More information about the 1929 searches for the *Southern Cross* and the *Kookaburra* may be gained by reading the Government report entitled, *Report of the Air Inquiry Committee in connection with the flight of aeroplanes 'Southern Cross' and 'Kookaburra', March — April 1929,* which is published by H.J. Green, Government Printer.

The funerals

During April 1929, the Federal Government decided that the bodies of Anderson and Hitchcock would be brought to Sydney and honoured with an appropriate ceremony. Mrs Violet Hitchcock objected, stating that she wished her husband to be buried in Perth. She contacted Tom Fox, the secretary of the Wharf Lumpers Union at Fremantle who telephoned John Curtin, the local Federal Member of Parliament. Curtin sent this cable to the Prime Minister:

'Mechanic Hitchcock's widow and their three children request Commonwealth Government arrange Hitchcock to be interred Perth Stop This to be at nation's expense Stop View of all circumstances associated tragedy believe your authorization this course would meet national approval and would be comfort to widow of brave Australian.'

Prime Minister Bruce agreed and arrangements were made for Anderson to be buried in Sydney and his mother to be brought there for the funeral. As requested, Bob Hitchcock would be buried in Perth.

As soon as the news was published, Motor Funerals Ltd sent a cable to the Defence Department offering to donate their services to bury both bodies anywhere in the Commonwealth. In offering to 'arrange all details', the firm said it required only to be covered for out-of-pocket expenses.

The offer was accepted.

Bob Hitchcock was buried first, in a modest civil ceremony on July 3rd, 1929 in Perth's Karrakatta Cemetery. As her husband had been a staunch member of the Church of Christ, Violet Hitchcock decided against a funeral with military honours, saying that she preferred a simple family funeral, which would be open to the public. Hitchcock was buried as he had lived for 37 years — quietly and unobtrusively. Hundreds of people came to pay their last respects.

At 9 am on June 30th, 1929 the Melbourne Express arrived in Sydney, carrying the casket of

By request, Hitchcock was accorded a quiet funeral in Perth. It was attended by members of his family. Two of his sons and several relatives are pictured here. His wife, Violet, had requested a minimum of publicity and did not pose for this photograph. Hitchcock was buried in Perth on July 24th, 1929.

Mr A.V. Hilliard, father of Anderson's fiancée. He also looked after her legal interests concerning the search and the recovery of the bodies.

Keith Anderson. Six officers of the Royal Australian Air Force walked to the end of the train and drew out the coffin. Bon Hilliard's father stepped forward with a wreath of lillies and sweet peas on his daughter's behalf. Beside him stood the Minister for Home Affairs, Mr Abbott, representing the Federal Government. With hands clad in leather gloves, the officers lifted the coffin to their shoulders and walked past the sombre gathering.

The coffin was taken by hearse to the mortuary chapel of Motor Funerals Ltd, in City Road, where it lay in state for several days. On July 5th, it was taken to Saint Stephen's Church in Macquarie Street, Sydney and one of the largest crowds ever to have assembled for such an occasion, visited the church. One city newspaper said:

'From early morning until nightfall, thousands filed before the coffin, the queue stretching as far as Philip Street. Reverently, they bowed their heads to the hero whose only thought was to fly to the aid of his mates. Some women knelt by the coffin and prayed, others sobbed.'

A rival newspaper resorted to more colourful prose:

'Mingled with dust and ashes — the red dust of the desert and the ashes of the fire he lit in vain to signal for aid — the mortal remains of Keith Anderson lay in the wasteland. Until civilization demanded its dead and men went out and faced hardship and danger to win its spoil from the wilderness. This they achieved and so in Saint Stephen's yesterday, the airman's casket lay in state.

'Sydney did homage, reverent, moved — and proud. Men and women of all ages, of all sorts, of all grades of wealth and poverty, walked past the coffin with its wreaths and its pall, honouring the hero that perished in sacrifice.

'He died in solitude — he has lain in state — may he Rest In Peace.'

The coffin was taken to the Presbyterian Church in Mosman, an inner city suburb, where thousands more people filed slowly past. Fifty men from the Returned Soldiers League kept vigil throughout the night. Early on Saturday afternoon, July 6th, the casket was carried from the church to a specially prepared grave, facing the sunrise. The ground, in Rawson Park at George's Heights, was donated by Mosman Council.

Six thousand people, including representatives from a large number of official bodies attended the funeral. Mr. A.V. Hilliard made himself available for Press photographs but his daughter, along with Mrs Anderson, Milton Kent and Jack Cantor were less conspicuous in their grief.

Selected soldiers, their medals glittering in the bright sunshine, fired their rifles in salute. An army bugler blew the thin, sad notes of the Last Post.

One Sydney newspaper described the funeral thus:

'As the coffin was born from Mosman Presbyterian Church to Rawson Park, the nation paid homage to the memory of a brave man. Governments and public bodies, war establishments and the institutions of peace were represented officially. Soldiers walked with arms reversed before the gun carriage which bore the airman's body; blue clad mechanics of the Air Force marched on either side; a Minister of the Commonwealth walked bare headed among the chief mourners;

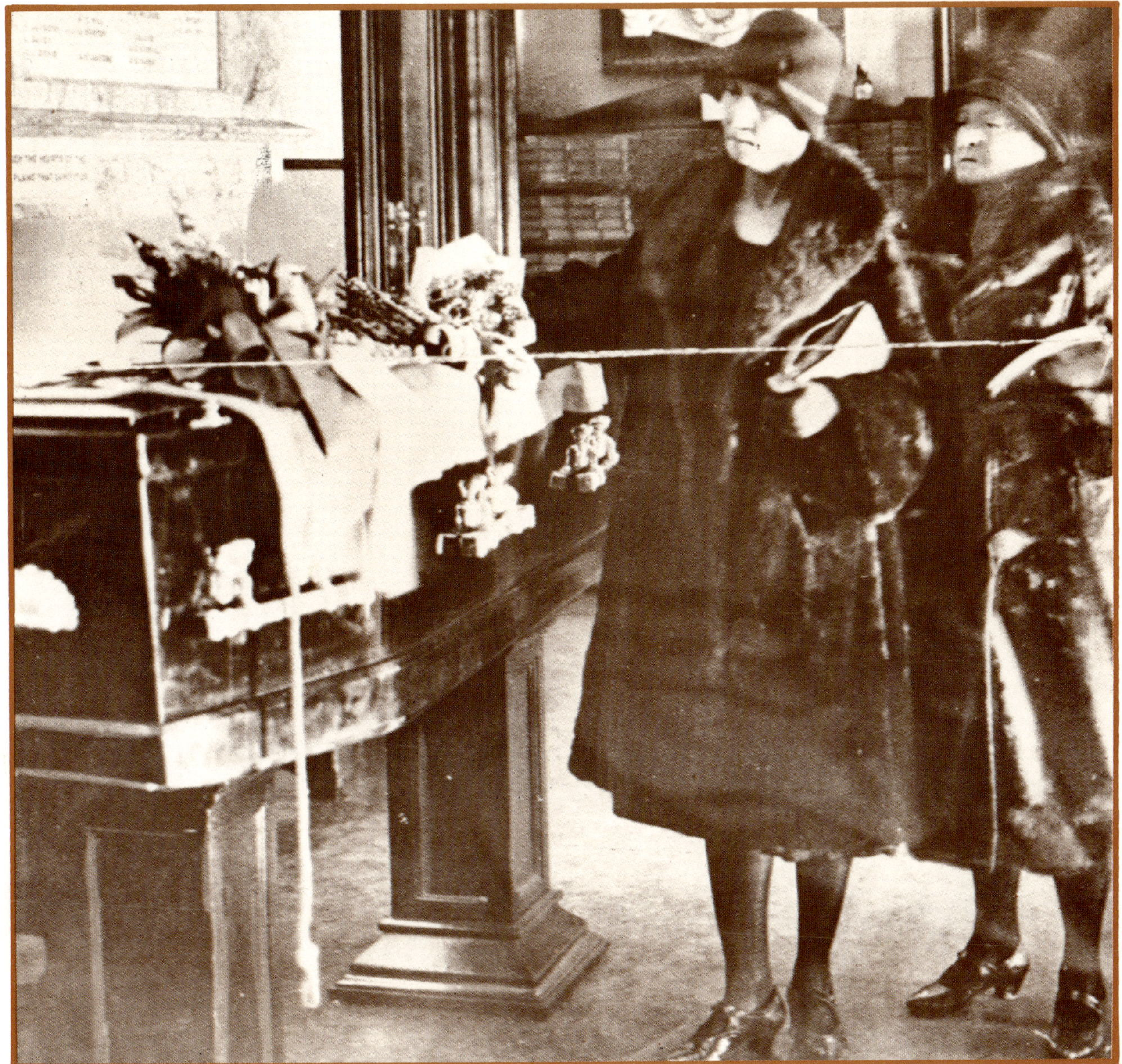

aeroplanes swooping down to near the level of the tree-tops dropped wreaths and flowers.

'But the most remarkable tribute, because it was spontaneous, came from the people. They gathered in thousands in the quiet suburban street before the church and in thousands following the cortége over the rough track to the high open space where the grave had been dug. There was no demonstration, no talking. They came simply and quietly to say farewell to one of those young Australians who have made Australian aviation famous for all time ...

'Very wisely, those in charge had fixed amplifiers on the front of the church so that the large audience could follow the proceedings within. The simple, very eloquent words of the Reverend D.P. Macdonald were clearly heard a hundred yards from the church...'

A touching moment came when five tiny Tiger Moths flew in formation above the burial site, forming a cross. They were followed by eight RAAF planes which flew in low to drop wreaths in tribute, then they turned slowly away.

A remarkable event had occurred earlier in the day. A kookaburra bird had flown to the top of the church's flag-pole and looked solemnly down. It stayed there for some hours and was observed to leave only after the coffin had been lowered into the ground. The next day, Mr Hilliard told the Press that no kookaburra had ever visited his home before, but on the morning of April 21st (the day Anderson's body was found), a kookaburra had settled on a tree opposite the house. Its laughter gave cheer to his

daughter, he said, but it turned out to be a sign of tragedy, not joy. Then the bird disappeared and did not return until the body of the dead aviator had come back from the desert.

Many fine words were spoken and public promises made during the emotional days leading to the burial scene. When the Federal Government had earlier announced that it would, at its own expense, transport the bodies to a suitable resting place, the Prime Minister added that the Government proposed to erect a memorial on the aerodrome at Alice Springs, which had been the first aerodrome on the recognized aerial route after it leaves the Overland Telegraph and would be named after the dead aviators. Neither of the promises was honoured.

However Anderson did achieve the permanent place in aviation history that he had always sought. His epitaph, engraved in stone at his burial site, reads:

Anderson was accorded a hero's funeral and one of the largest crowds ever seen assembled to see the official cortége at Mosman, Sydney.

The final words were read, as Anderson was buried at George's Heights, Sydney in ground donated by the Mosman Council. Six thousand people attended the ceremony.

'The passing years shall leave him ageless, loved and unforgot. In affectionate remembrance of Flight-Lieut. Keith Anderson, late 73rd Squadron, Royal Flying Corps.

'A hero and a very gallant gentleman who, whilst searching in his mono-plane the *Kookaburra* with his mechanic H.S. Hitchcock for the lost *Southern Cross* perished from thirst in the Australian desert 12th April, 1929 and was here buried 6th July, 1929.

'Greater Love Hath No Man Than This.'

The monument was built, not by the Government, but by public subscription from the people of New South Wales and it was officially dedicated by the Governor of New South Wales, Sir Philip Game.

John Fairfax and Sons

The search for desert gold

Without question, gold lies in the Tanami Desert. Some of the first white men to endure the desert's harsh hospitality returned with yellow wealth. In 1913 gold-fever raged in the tiny township of Tanami when a new seam was found. So a crude mine was built in desolate terrain.

A London firm heard of the find and commissioned Bob Robinson, a West Australian mining engineer, to travel to the site and make an assay. Accompanied by two colleagues, Robinson undertook the first recorded motor journey into the desert, travelling from Wyndham to Tanami in a Model T Ford. The sheer isolation of the country was emphasized when the party eventually reached its goal and sent a 'telegraph' message home, advising of their safe arrival. It was carried by the first available camel-train, which left five days after Robinson had reached the gold-mine! Some weeks later the assay report was sent to his London principals but details were not made public.

The gold-mining boom which gripped several states during the latter part of the last century gave prevalence to claims and counter-claims about gold discoveries, but not all of them caused great enthusiasm or action. During the late 1890s, reports were rife that a legendary gold-reef existed somewhere in the desert in Western Australia. The stories were based on a report that a skeleton had been found clutching a large bag of gold. For the next 40 years adventurers from all parts of the world came to seek the fabulous reef. Pilot Lester Brain for example, was hired to make aerial searches for gold a long time before the *Kookaburra* became lost. During the mid-1920s, he had flown American, Jim Stark over thousands of kilometres of parched terrain in search of a gold-reef.

In a letter written to Mr Max Morris of Adelaide in 1973, Brain said:
'Details of where I flew with Jim Stark have never been published. The latitude and longitude point where he expected to rediscover a reputed rich gold-reef was very close to where the *Kookaburra* went down. But in our search — 1925/6 — we were unable to find anything which looked like a gold-reef.

'Whilst on the first of two trips with me, Jim Stark was definite that the area was the locality of his legendary gold-reef. He returned to America and later came back to Australia and continued his search with ground parties some hundreds of miles further south! As far as I am aware, he never did find what he was looking for.'

At the time of those searches, the name Lasseter's Reef had not yet been coined, but it is likely that Stark — and numerous others — were looking for the same reef. Searches are still continuing today. To some, discovery of 'The Reef' holds the promise of untold wealth just waiting for a lucky person; to others, it is a myth built upon a fantasy with as much basis as the fabled pot at the end of the rainbow.

Harold Bell Lasseter was 17 years old around the turn of the century, when he attempted to walk alone from Alice Springs to the West Australian coast. His adventure would probably have cost him his life, had he not been found by a surveyor named Harding. Lasseter told Harding that whilst lost in the region between Ehrenberg and the Petermann Ranges, he had found a huge gold-reef, some 10 miles (16 kilometres) long.

Although he told his story on many occasions, no action was taken. Some years later, Lasseter claimed that in 1903, he and Harding had returned to the area and located the reef with great difficulty. He said that West Australia's current gold-boom, and the fact that their assay showed only three ounces (85 grams) to the ton, discouraged anyone from supporting a full-scale venture. Lasseter and Harding drifted apart shortly before Harding died.

It was not until 1930 that Lasseter appeared at the Sydney headquarters of the Australian Workers Union and his story received publicity. He was able to persuade a group of speculators to form the Central Australian Gold Expedition Company Ltd, with capital of £5,000, a very considerable sum at the time. The company decided to mount an expedition with Fred Blakely as leader and Lasseter as guide.

The expedition included a Thornycroft six-wheeled truck, (which had previously been used to collect the bodies from the site where the *Kookaburra* had been forced to land), a team of camels and a light aircraft, the *Golden Quest*, a De Havilland Moth, which later crashed. A second aircraft named *Golden Quest II* was flown in. Unfortunately, this aircraft crashed near Ayers Rock, and Errol Coote, the pilot, spent several anxious days before being rescued.

Robert Lasseter

Harold Bell Lasseter claimed to have found a gold-reef around the turn of the century. This was taken when Lasseter was in his late 20s.

The expedition leader, Blakely later wrote an account of the search in his book, *Dream Millions.* 'Everyone was riding on a wave of enthusiasm,' he wrote, 'for here was something the country really needed. Harry Lasseter, the little nuggety man, seemed to have shown a way to break the depression. The patriotic spirit was really grand ...'

Very quickly, Blakely and his colleagues learned that Lasseter was not the skilled bushman whom they had expected. As the search continued, some began to doubt his story, but Lasseter never lost his own faith. When the expedition decided to return to Sydney, he insisted on continuing by himself, with two camels to carry his supplies. He was never seen again.

The company hired a skilled Northern Territory cattleman, Bob Buck, to search for the missing man. Some time later, Buck returned, claiming to have found Lasseter's body on a dry river bed near the Petermann Ranges. He said that he had buried it there. Buck sent the company some notes which he said he had found on Lasseter, together with a photograph of the grave. A death certificate was then issued at Alice Springs, giving the date of death as January 30th, 1931.

Followers of Lasseter lore believe that Buck may have concocted the story or found the wrong man. In the latter case, he would also have fabricated the notes, which claimed that Lasseter's camels had bolted but he had managed to relocate the reef and peg it out. (It was not explained how this was done without the camels and tool pack.) The notes also said that Lasseter had become lost again and had been found by Aboriginals shortly before his death.

According to Harry Purvis, a celebrated airman who was later hired to help find Lasseter's Reef by means of an aerial search, some human remains which were found during the 1960s near the Docker River were definitely established to be those of Harold Bell Lasseter. They were buried at a cemetary in Alice Springs, with a sculpture of a prospector with a wide hat and a gold-pan.

Regardless of the truth of the Lasseter story, the legend lived on. No doubt it was fed by Ion Idriess' book, *Lasseter's Last Ride* (1931). Some believe that the book was a romanticized account of the search based on the notes that Bob Buck claimed to have found. In his book, *Dream Millions,* Fred Blakely concluded that Lasseter had deluded himself; he had told his story so often, that he eventually came to believe it. Whether this is true or not, this judgement did little to discourage the gold-seekers. At least 22 full-scale expeditions have since been mounted to find the legendary gold-reef.

In many ways, a strong similarity exists between the legend of Lasseter's Reef and the story of the *Kookaburra.* Both were surrounded by rumours and their locations, elusive. Despite the brave words spoken at Anderson's funeral, no serious attempt was made to reclaim the plane after the Thornycroft expedition had cleared a suitable runway to allow it to be flown out. Lack of money was the probable reason. Even a rudimentary examination of the problems showed that the recovery cost would far exceed the plane's commercial value.

However dozens of searches were made to relocate it in later years and many of them were motivated by the lure of gold. The story of a rich gold-bearing reef existing only 22 kilometres from the abandoned *Kookaburra* became widely known, and even today it is uncertain whether the claim was born in fact or fantasy.

It started when Frank Nottle announced that when returning from the Thornycroft expedition he had picked up a rock which was later found to have an unusually high gold content. His account held some contradictions but these may have resulted from the long delay between his return from the *Kookaburra* and the 'announcement' of the gold discovery. His convincing tale persuaded several level-headed investors to spend thousands of dollars looking for the *Kookaburra* and the gold-reef which they believed to be nearby.

Meanwhile, the longer the plane remained in the desert, the more difficult became the recovery. As the patches of burned scrub regained their natural colour, location became increasingly difficult and soon the challenge was to relocate the plane, and not necessarily to extract it from the Tanami Desert.

It was not until 1961 that someone actually saw the remains of the *Kookaburra* again — although dozens of attempts had been made.

Two enthusiastic supporters of Nottle's story were T. Granton Harrison and Rod Liebernecht. Between them they spent sums totalling the then-equivalent of $40,000 trying to locate the gold-reef. Their reasons for believing Frank Nottle's story were given to journalist Frank Kennedy and published in *Dominion*, the Auckland newspaper. The article appeared in September 1975, just before Liebernecht departed on his ninth expedition in

This photograph gives some indication of the density of flies in the desert. The man is Tim Hunt who was a member of Liebernecht's expedition in November, 1976.

search of the *Kookaburra*. Kennedy quoted Liebernecht as saying:

'Frank, a keen rock hound, only had time to hammer a specimen, the size of a man's two fists, from an outcrop while his party mended one of the many punctures whilst transporting the two airmen's bodies from the *Kookaburra* to Newcastle Waters in lead-lined coffins.

'Then the party called, "Come on Frank, we are ready to go".

'The reason for the hurry was that the expedition travelling in a six-wheeled Thornycroft truck, was running short of water and fuel.

'So Frank, with no time for proper study of the rock or the environment, possibly lost the chance of becoming a millionaire when he casually threw the specimen in the back of the truck.

'Many weeks later, after having brought the airmen's bodies to Adelaide, Frank at last broke up the rock specimen. He was astounded to find the pieces richly flecked with gold, a richness confirmed by assay.

'Onset of the Great Depression, precipitated in 1929 by the New York Wall Street stock market crash, meant years of delay before Frank, then Thornycroft's South Australian manager, could mount the first of his four expeditions to search for the gold-crop.

'By then his recollections had dimmed, and he never found it again.

'However, before his death, Frank encouraged me to persist.'

Liebernecht went on to describe the harshness of the desert and the problems associated with the search, adding:

'I had 44 punctures within six and a half days when I reached within two miles* of the *Kookaburra* in June 1968.'

Others who have studied the *Kookaburra* story and the prospect of finding gold have added to Frank Nottle's story.

South Australian Max Harris has organized his own expeditions to find the gold-reef and in 1977, he wrote to researcher Dick Smith suggesting a joint gold-seeking venture. Harris says that Nottle gave one of the rock samples to a helper when unloading the Thornycroft and while examining the other piece, he detected a trace of gold and broke it open. According to Harris, Nottle then had an assay done by the late Sir Douglas Mawson which reputedly showed the quartz held between 60 to 90 ounces (1.7 to 2.5 kilograms) of gold per ton. In his letter, Harris told Dick Smith that he had seen and photographed the gold-reef from the air, and therefore knew its exact location, including

** 3.2 kilometres, this was in fact an approximation by Liebernecht.*

bearings, grid references and distances from land marks.

Like so many before him, Harris failed to locate the *Kookaburra* itself, the site of which was generally believed to hold the key to the alleged gold-reef.

His searches added to many other attempts which had been made since 1930 to find the *Kookaburra* and/or the gold. Another was organized by the Perth Museum, which holds the tail fabric on which Anderson wrote his diary. According to Vernon T. O'Brien of Darwin, some Aboriginals once set out to find the *Kookaburra* with the idea of installing it in a shed as a tourist attraction at Hooker Creek!

It was not until 1961 that the *Kookaburra* was seen again. A party led by Vern O'Brien was about to cross 320 kilometres of desert from Wave Hill station to Powell Creek on the Stuart Highway. They were investigating the proposed plan for a pastoral road. One hundred and forty-nine kilometres from the point where they had turned east from Wave Hill — Hooker Creek Road — they stumbled across the remains of the *Kookaburra*.

O'Brien expected to find a body in the plane but the cockpit was empty. He photographed the remains and removed the engine number identification plate and instrument panel, along with the smashed compass. When the wreckage was later identified as the *Kookaburra,* his curiosity was aroused. O'Brien took part in two subsequent but unsuccessful searches to relocate the plane.

In 1972, at the request of the then-administrator of the Northern Territory, an attempt was made to relocate the remains of the *Kookaburra.* The aerial search was conducted by the RAAF during normal flying operations. Vern O'Brien acted as a consultant.

O'Brien participated in another search in 1974 but this was also unsuccessful.

John Haslett, the honorary curator of the aviation section of Darwin's Military Museum, has conducted two searches. An avid historian and a most knowledgeable man on early aviation, his object was to recover the *Kookaburra* and then display it in the museum. He envisaged a striking display with a background showing the authentic desert setting in which the airmen had perished. The first attempt in 1971 was an aerial search made with a private aircraft and in the second in 1974, Haslett led five people with three vehicles into the desert. Both attempts failed since they were

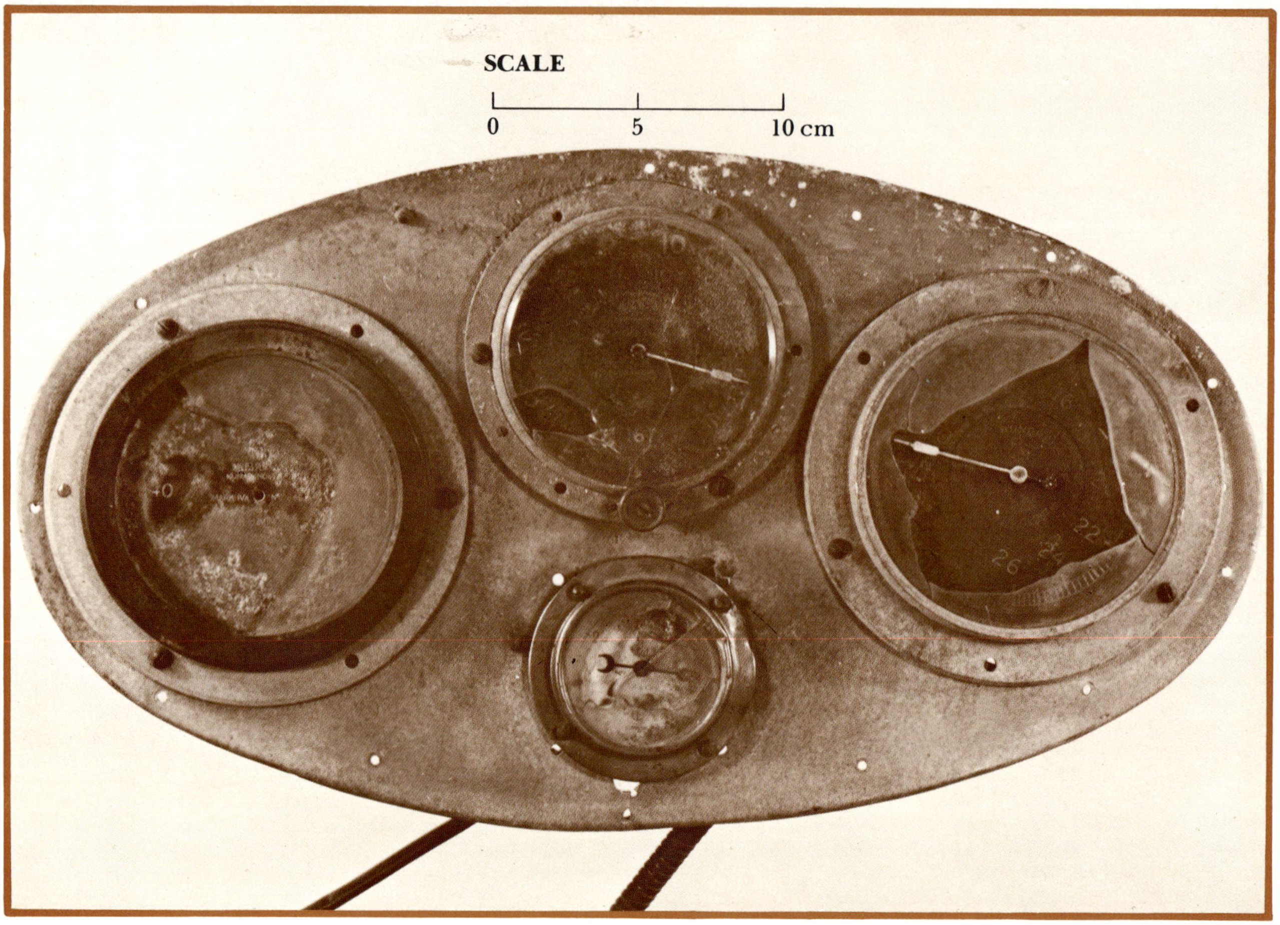

The instrument panel which was removed by Vern O'Brien when he accidentally discovered the *Kookaburra* in the Tanami Desert.

The three vehicles used in John Haslett's search in August, 1974. The men stopped to fill their drinking water tanks at Katherine River en route to Wave Hill. Note the freezer cabinet on the back of the trailer.

inadequately financed and relied on dead reckoning for their bearings.

In 1975 the first official search was mounted since the Thornycroft expedition. Organized by Dr Colin Jack-Hinton, a director of the Northern Territory Art Galleries and Museum Board, it was backed by Federal funds. Vern O'Brien and John Haslett joined the party. The expedition was accompanied by an ABC camera crew and by Gary Moseley, a Vietnam War veteran, who piloted the expedition's helicopter.

For 10 days the party crashed through thick scrub and, aided by the helicopter, they searched the area where Vern O'Brien believed that the *Kookaburra* lay. Despite his enormous experience as a surveyor and his previous relocation of the wreckage, O'Brien was unable to find any trace of it. Ironically the final, successful expedition to locate the *Kookaburra* found wheel marks, which indicated that the ABC truck had come within a few hundred metres of the plane. Nevertheless, the ABC crew, directed by David Poynter, made an interesting documentary for the television programme *A Big Country*.

A year later, in April 1976, the acting administrator of the Northern Territory had the remains of the *Kookaburra* classified and protected under a special ordinance so as to

Members of John Haslett's search for the *Kookaburra* in 1974, erected a water catchment drum and a navigation mast, left, at the centre of the search area. This position was based on Vern O'Brien's estimate of the location of the *Kookaburra*. The water catchment drum, above, recorded the date, the names of the six men in the expedition and the approximate position of the centre of their search area.

prevent the removal of any part overseas or interstate.

Although searches for the gold-reef are still continuing as this book is being written, the man who eventually relocated the *Kookaburra* — researcher Dick Smith — believes that the alleged existence of gold is a myth. Having volunteered the exact location of the *Kookaburra* site, he has publicly stated that he has no plans to go in search of gold. The most sophisticated techniques possible, including aerial surveys, he says, have yet to discover gold-bearing rock in the area.

Dick Smith's searches

Dick Smith is a man of many facets. His vitality has helped him to become a millionaire by his 30th birthday and, he can now afford to indulge his whims. Dick's interests extend from farming to mountain climbing, from family life to his multi-million-dollar electronic business.

However second only to his wife Pip and their two daughters, is his enthusiasm of aviation. Dick has owned several aircraft, and regularly commutes in his Bell Jet helicopter between his home in a rural suburb and his office in North Ryde, Sydney.

His business centres around the manufacture, importation and distribution of electronic equipment and components. In 10 years it has grown from a single shop front to a $15-million a year conglomerate with 11 Dick Smith stores, plus 160 distributors around the country.

With his extensive resources and his in-imitable tenacity, Dick Smith went in search of the *Kookaburra*. He considered transporting it back to Sydney — where its fatal journey had started — and restoring it for permanent exhibition in the international section of Kingsford Smith airport.

'My obsession — for that is what it became — with finding the *Kookaburra* was inevitably going to invite the sort of accusations that all too often occur,' he says.

'Nobody thinks you are "fair dinkum". They believe you must be concerned with personal publicity, glory or trying to further your business interests.

'The fact is that I wanted to solve a mystery. Although I had always been fascinated by aviation history and its pioneers, I did not realize that the *Kookaburra* had never been recovered.

'After compiling some research on the subject, I concluded that the searchers to date had not really known where to start, nor did they know their precise location whilst actually searching.'

In April 1977, Dick Smith flew his Twin Comanche to Alice Springs and followed the exact route that Anderson and Hitchcock had taken in 1929.

'Even in this comfortable and sophisticated aircraft,' he says, 'it was a chilling experience to stare down at that appalling wasteland which they had tried to cross in their tiny old-fashioned and patched-up mono-plane.'

During the flight he noticed a curious depression in the terrain but attached no special importance to it at the time. It would later prove to be a key factor in finding the plane.

When Dick reached Wave Hill station and met the manager, Ralph Hayes, he noticed a helicopter which was used there for cattle mustering. He hired it and flew to where the *Kookaburra* was believed to be. Fifty years

The first Dick Smith expedition leaving Wave Hill in 1977.

earlier, the search would have been relatively easy. The pilot would have headed for the large area of burned bush and looked for silver wings glinting in the sunlight. But the bush had regrown and several fires, together with the ravages of the harsh climate, had reduced the plane to a wreck, with the canvas burned off the wings and fuselage.

Dick did not expect to locate the plane from the helicopter, but he wanted to 'get the feel' of the area and gain experience of some of the problems which had caused so many search expeditions to fail.

After travelling for about 128 kilometres in a south-easterly direction from Wave Hill, the helicopter landed. The area was surrounded by turpentine scrub several metres high. After taking a few steps, Dick turned around and discovered that he could no longer see the helicopter.

'For one brief and terrifying moment, I was lost,' he recalls. 'I realized for the first time the magnitude of the task. To find the *Kookaburra,* I would need to organize a sophisticated expedition.'

Plans were laid as soon as he returned to Sydney, with a target date set for July 1977. Dick pin-pointed accurate navigation as the key to success. He started thinking about the need for a surveyor who could work with theodolites and aerial maps and realized that such a man (Nigel Davies) was already on the staff of his own company. He advertised for interested parties and, from 80 serious replies, he selected 18 volunteers, including David Poynter who had directed the ABC television documentary.

In addition to the personnel, Dick organized six land vehicles, ranging from a giant Mercedes Unimog to a diminutive Suzuki four-wheel-drive utility. There was also a helicopter, a

Nigel and Dick using the theodolite to accurately fix the position of Anderson's Corner, latitude 18°S and longitude 132°E.

fixed-wing aircraft and a large earth-grader. As the planning proceeded, it became obvious that the logistics would be enormous as they involved fuel, possible vehicle breakdowns and supplies for the 18-member expedition.

The cavalcade of road transport left Sydney on June 25th, 1977. Dick left a week later in a Piper Chieftain, with other members of the team.

A minor problem occurred with regard to supplies. Each member of the expedition would be responsible for their own supplies, but some had not experienced life in the bush before. Gerry Nolan (the expedition co-ordinator) had supplemented the food supplies with ingredients such as potatoes, pumpkins and tins of fruit but no one had thought to include extra quantities of beer. Since beer must be transported a long distance, the price of a can in outback towns is about four times the usual price. Dick said that at one point, 'I sent an expedition member back with the pilot to Tennant Creek to buy more beer'.

'The main thing on my mind, however, was not beer or supplies, but where to begin. Turning the point of location over and over in my mind, I became increasingly convinced that Vern O'Brien's estimate, made in 1961, was not too accurate. However, it was obviously the best starting point for our search.'

So this point — which was at the intersection of latitude 18 degrees and longitude 132 degrees east — was dubbed 'Anderson's Corner'.

Another idea was suggested by Gerry Nolan. He had telephoned Dick from Wave Hill to say that the station's huge grader would be ideal for cutting a road to Anderson's Corner and that the manager was agreeable to hiring it for that purpose as he wished to have such a road for fire-fighting purposes. Dick agreed to the extra cost.

Gerry directed the road-making task, the grader being driven by a magnificent old Territorian named Claypan. He made no secret of his amusement at the apparently crazy idea, since he knew how quickly nature would reclaim her own. He was however extremely experienced and helpful, clearing the road in only two days. During the work, the huge tyres on the grader were punctured 17 times and Dick commented on the pitifully thin tyres fitted onto the *Kookaburra*.

'What sort of chances did it have against the turpentine?' he asked.

The road shortened the journey from Wave Hill to Anderson's Corner from an estimated four days to two and a half hours, and the idea proved to be one of the expedition's luckier breaks. The time saved thus made it possible to extend the actual search to five full days. Cost was not the key factor since the helicopter alone was $130 per hour.

Nigel Davies had taken astro fixes (star shots) and plotted a series of grids on which the search would be based. Dick and his associates had agreed that the key to success lay in accurate survey work, complemented by a thorough search. This could only be assured by dividing the area into relatively small 'parcels' and scrutinizing each one in turn.

The base camp was established at latitude 17 degrees 59 minutes, 55 seconds and longitude 132 degrees, 28 seconds. In other words, the expedition camped at a point 950 metres east of Anderson's Corner and 150 metres to its north, an error of only one kilometre. For the rest of the search, the astro fixes were taken constantly so that at all times, the members of the expedition knew precisely where they were.

The grid search began in earnest on the Wednesday and proved more complex and difficult than had been expected. On large scale maps they had marked out a series of rectangles, each measuring 3.2 by 1.6 kilometres. The plan called for the helicopter to place a man with an aluminium marker pole carrying a coloured flag, on the corner of each rectangle.

The men establishing the grid corners on the ground would leave trails of coloured tape behind them to be sure that they would easily be able to retrace their steps as they paced out the rectangle.

The huge tyres of the grader were pierced many times by the sharp turpentine scrub, causing the expedition to stop and repair the punctures

A Bell 47 helicopter was used during both of Dick Smith's searches for the *Kookaburra*, because of its flexible flight plan and ability to land in rough terrain.

As each man covered 400 metres, the helicopter, with an observor, flew overhead scrutinizing the terrain below. It proved tedious, grinding work, especially for Bob Coombe the pilot, who flew from eight to 10 hours each day. It was however the only way to be sure of finding the plane in the tangled turpentine scrub.

One grid search was done eastwards, over the same ground that had been covered by the expedition mounted by the Northern Territory Museum Board in 1975. The party then worked westwards and finally southwards, painstakingly searching each grid on their map.

As the eye-aching search proceeded, disaster struck.

'It was Thursday, July 7th,' says Dick. 'I had expected the helicopter to return no later than noon for refuelling, but it did not appear. Bob Coombe knew and understood the country. He was also perfectly familiar with his machine, so after two hours of anxious waiting, I radioed the Flying Doctor Service at Alice Springs. They put me through to Darwin and the Department of Civil Aviation, who declared a full-scale emergency. They said that all aircraft in the area would be diverted to look for the helicopter.

'Soon after three o'clock, the pilot, observer and two grid markers came stumbling out of the bush. I was relieved to see them and promptly called off the search. One of the grid markers had accidentally got his marking pole caught in the spinning helicopter blades. No one was hurt but the helicopter was now out of action and the enterprise was literally grounded.'

Dick sat down to review the situation and concluded that the enterprise had become too complicated. With only one day left of the scheduled five-day search, and still no signs of success, there was little to do but quit.

Further reflection convinced Dick that he did not wish to be just one more man who had searched in vain for the historic aircraft. The pitiless desert was not going to win.

The expedition's dying hours were described by an expedition member thus:
'At dusk, Dick laid the Anderson Memorial Plaque (he had brought two, one to be used if the plane had been found and another if he failed), under the large and beautiful **ghost gum**, which stood at Anderson's Corner, to commemorate the Dick Smith Expedition. He also named the road built by Claypan, the Hitchcock Highway. Dinner that night consisted of steak and hot potatoes. Next morning, Friday, we placed a 44-gallon* drum of water under the ghost gum and left for Wave Hill. We loaded the aircraft there and departed at 1.40 pm.'

The purpose of the plaque was to acknowledge the bravery of Anderson and Hitchcock and the rigours of the treacherous country. Dick also addressed the expedition, stressing that he would return.

'We had learned, even in this short time, about the meaning of mateship which had meant so much to the dead aviators,' he said later.

'I was certain I would come back to try again but as we were boarding the plane at Wave Hill, I caught a look from Claypan. He had seen, I suppose, many people come and go in search of the elusive *Kookaburra*. He gave a wry, almost

200 litres.

122   ironic grin, as if to say, "Well, it's mighty tough country".'

On his return to Sydney, Dick sent details of the search to the Northern Territory authorities then upon learning that Lester Brain was still alive Dick telephoned him, hoping that some additional information would provide the breakthrough. He found Lester Brain was reluctant to discuss the subject in detail. He took the view, with some justification, that so much nonsense had already been written about Anderson and Hitchcock and that no one should stir the story.

However, the reticence was short-lived. The two men soon met and the legendary pilot recalled the 49-year-old incident as clearly as one would relate a recent event. He told Dick about the blow-hole* which he had noticed on his way back to Wave Hill in 1929, after passing over the *Kookaburra*. Dick's interest grew. Perhaps this was the clue; the way to find the plane was to locate the blow-hole and work from there. Possibly it was the depression which he had already seen during his Comanche flight to Wave Hill station, the previous year.

Although Dick immersed himself in his electronics business, he could not forget the *Kookaburra*. His decision was finally made to try again when his service manager, Garry Crapp, started to 'needle' Dick about letting the desert win.

'Look Garry,' he retorted. 'If you are prepared to tow my caravan all the way to the desert, so I can have my family with me this time, I will go.'

** A blow-hole is caused by a local collapse of the earth's crust.*

With the commitment made, Dick commenced plans and selected people for the expedition. He decided that the blow-hole provided the only dependable datum point on which the bearings could be made.

Aided by colleague Tony Peter, Garry Crapp headed north with the caravan, while Dick and Pip and their girls, Jenny and Hayley, travelled in their Beech Baron aircraft, with Noel Pottie, who is an experienced bushman. The party also included Keith Locke from Alice Springs.

Garry Crapp made arrangements to hire a truck at Katherine and to recruit Bob Coombe to pilot the helicopter. Dick contemplated using radar which, in theory, would be able to pick up the metal parts of the *Kookaburra* in the vast desert. This time the cost was prohibitive.

After lunching at Tennant Creek on Saturday August 26th, 1978, Dick and his party flew to Wave Hill, and tried en route to find the blow-hole. Pip Smith spotted it at about 3 pm and, even from 900 metres, it appeared as a great gaping hole. After circling to make a thorough examination, they estimated that it was about eight kilometres from Hitchcock Highway. Greatly excited, the group flew to Wave Hill.

Next day they set off and, on Sunday August 27th, they took the grader right into Anderson's Corner.

'On the Monday, we put in two roads,' says Dick. 'One to the east and one to the west from Anderson's Corner. We were still planning to search in the Vern O'Brien area, which was basically around Anderson's Corner. We used the east and west roads, which ran for six kilometres each way, as a place to stand with our mirrors to get the helicopter run-offs.

'We worked on that until Monday as the helicopter was due on the Tuesday. The first task, when it arrived, was to establish if the

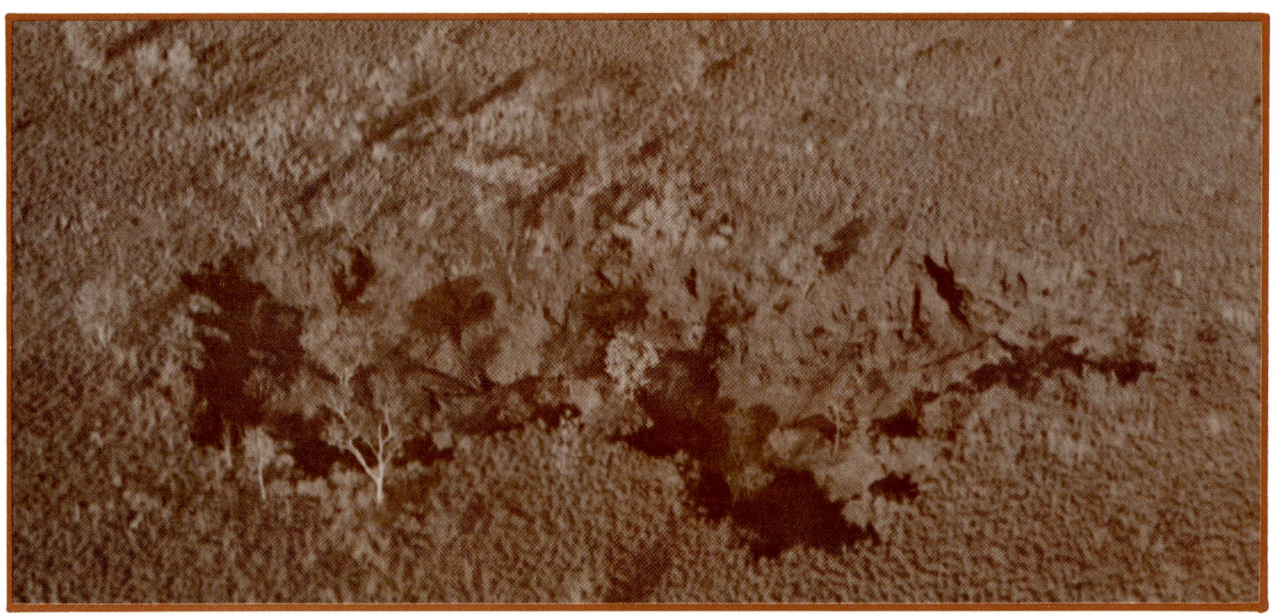

blow-hole I had seen from my Beech Baron was the correct one. Pip, myself and the pilot, Bob Coombe, flew north. After about half an hour we managed to locate the blow-hole and land.

'It was a most incredible thing, a huge hole in the ground. I said to Pip, "Look, if this is the right blow-hole — and this is the only one we've seen which looks anything like it — there should be some evidence of the 1929 land party". Pip found a number of tin cans which had been opened by pen-knife, and established that this was obviously where the party had been. We could see the actual place where Sergeant Douglas had fallen.

'We went back in the helicopter and started searching. We didn't think at the time to measure the distance from the blow-hole to Anderson's Corner, and we searched all day in the area where Vern O'Brien said he had found the plane. That night, when we were sitting in the little caravan, Bob Coombe managed to pinpoint the blow-hole on the aerial photos that we had. This was incredible, as the photos had been taken from 6,000 metres. We scaled off the aerial photos and marked the distance which Lester Brain said he had gone by air when flying from the *Kookaburra* over the blow-hole to Wave Hill station.

'Of course we knew that his air distance would have been just a very rough estimate, something like 29 kilometres. Suddenly Bob said:
"What a pity that we can't put a DME (distance measuring station) to the blow-hole." I replied, "That would be useless, as we don't have a mileage." Then, Bingo! I shouted, "We've found it".

'The others looked like I had gone a bit funny, but I had suddenly realized that we had the vital clue. We knew that the truck party's diary was inaccurate as far as directions went but of course the Thornycroft's odometer distances given in the diary, would be accurate.

'Kicking myself for not having thought of it before, I said, "Draw a line from the old Wave Hill station, over the blow-hole and straight out. The truck's odometer had shown 23 miles* and if we scale this off along the line, it will show us where the *Kookaburra* is. In other words, we should combine Lester Brain's directions with the truck odometer distance readings."

'We drew the line 23 miles out, and allowed a mile for probable wheel slip. That put the theoretical site of the *Kookaburra* about five or six miles north of where we were looking.

'I was very excited, but some of the others weren't. They thought it was crazy to go on information that was nearly 50 years old, and

Dick Smith with wife, Pip and daughters, Jenny and Hayley, who were part of the successful search party in August 1978.

that we would be better off following the more recent directions given by Vern O'Brien.

'But I wanted to go on the old information, because I have always felt that Lester Brain was an extremely good pilot and his directions would have been right. After a lot of discussion, this is what we agreed to do.

'On Wednesday we got up early and, with considerable enthusiasm, worked all day, searching this new area. Once again the search failed. We had virtually used up all our fuel and water, with nothing to show for it. And I felt terrible. We worked next day, Thursday, until lunch time but still didn't find anything. The helicopter came back with the news that we were almost out of fuel, and we all felt miserable. It seemed that the desert had won again.

'Almost in desperation I said, "I'll go up in the helicopter". I hadn't been up at all because my eyes are not as good as the others, and I'd been doing the ground marking with the mirrors.*

'The helicopter whirred into the sky with Tony Peter and I as observers. I told Bob Coombe, "Go further north because the truck might not have gone the full 22 or 23 miles, there might be more wheel slip than we thought. The truck was following the horse party tracks, and maybe they were not straight."

'Suddenly,' recalls Dick, 'quite clearly, there was the *Kookaburra*! Bob Coombe pressed the radio button and we all cried, "We've found it!

---

* *36.8 kilometres.*

* *By standing five or six kilometres apart and using the mirrors, they could reflect the sun to the helicopter to give them a point to which they could fly, thus keeping a straight flight path over the desert where there are no good land marks.*

The *Kookaburra* as Dick Smith and Tony Peter first saw it from the helicopter. The plane was barely visible since the turpentine scrub had regrown and the plane itself had been ravaged by the effects of the harsh weather and bush-fires.

We've found it!'' We were almost incoherent with excitement. It was a moment to remember forever, a time to dream about, to talk and laugh and cry about. The scrub was too heavy to land the helicopter, so at about 4.5 metres Tony Peter and I jumped out.

'It was exactly as Lester Brain had said, in line with Wave Hill station and the blow-hole and 21½ miles* from the blow-hole.

'Bob Coombe threw down a can of water and went back to ferry the rest of the party to us. As

* *34.4 kilometres.*

the helicopter disappeared, I again had that awful feeling experienced on my first helicopter trip from Wave Hill. That was when I walked into the scrub, could not see the helicopter and felt totally lost. I experienced a little of the loneliness and terror that Hitchcock and Anderson must have suffered.'

Dick looked around the *Kookaburra* and began to wonder how the aviators had ever managed to clear even a short runway with their bare hands and a pen-knife. He tugged at some of the turpentine scrub and realized how totally exhausting it would be to clear even a small patch.

He was surprised to find that, after the initial sighting of the plane, there was no sense of triumph. Instead, he was moved — almost overwhelmed — by sheer pathos. His throat tightened as he looked at the tiny plane, its skeleton ravaged by time, bush-fires and the

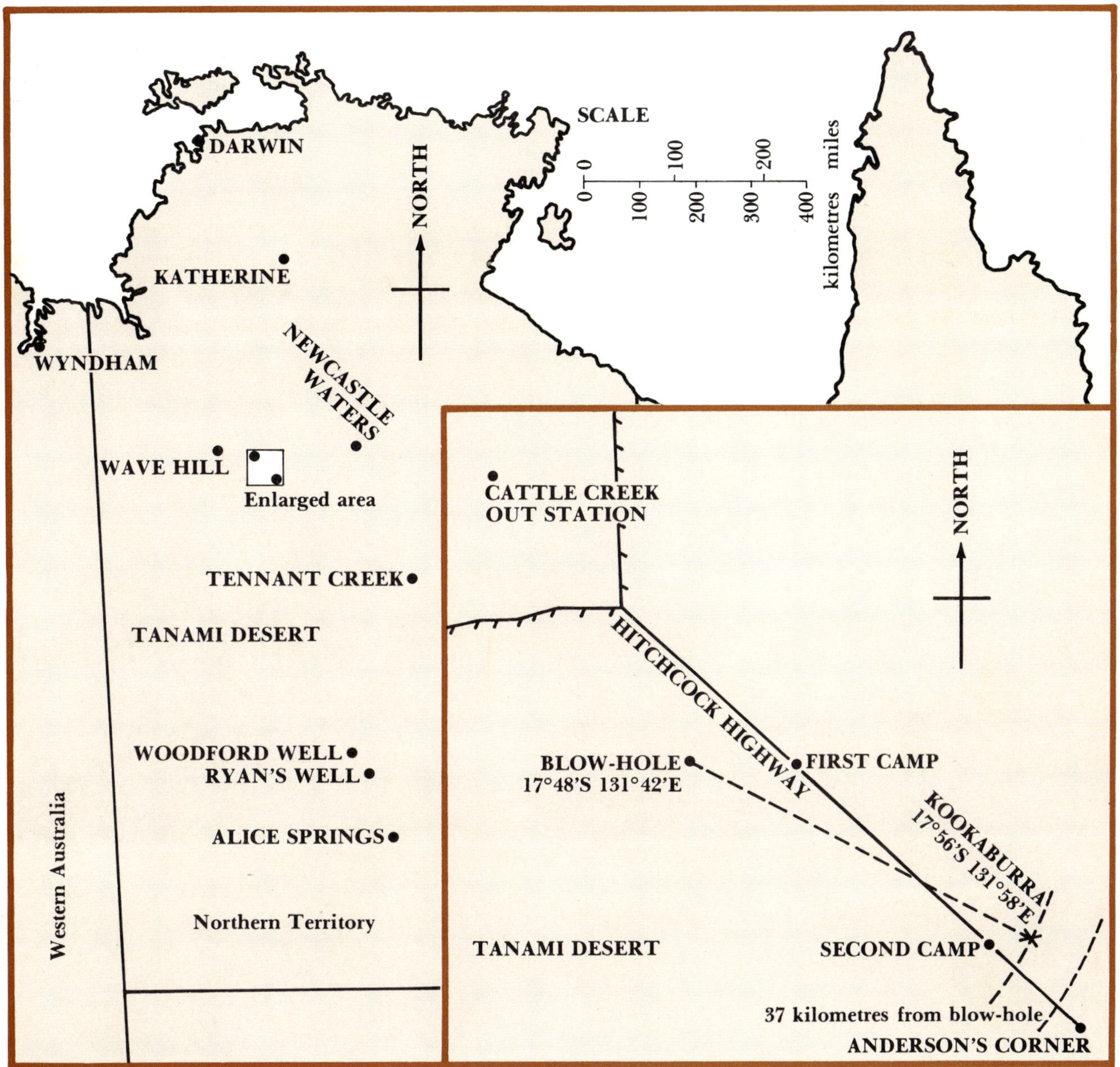

Map shows how Dick Smith located the *Kookaburra*. Using the information that Lester Brain had flown over a blow-hole on his way back to Wave Hill in 1929, and the Thornycroft truck party's recorded distances after they had spent the night at the blow-hole, Dick Smith drew a straight line and calculated the approximate position of the plane. The inset square shows the position of the search area in the Northern Territory.

Right, Dick Smith was deeply saddened by the eventual discovery of the *Kookaburra*.

relentless sun. How could men have ever dared such odds in such a small frail piece of equipment, he wondered.

Gripped by a sense of infinite sadness, he spun the rusted wheel on the undercarriage and looked at the faulty push-rod which had forced

This is how the *Kookaburra* looked when it was finally discovered by Dick Smith and his expedition, pictured right, in August, 1978.

Anderson to land in the desert. The propeller was burned off, but tiny brass screws remained. He could see where Vern O'Brien had removed the engine identification plate and instrument panel. Dick tried to picture Anderson and Hitchcock desperately trying to clear the runway 49 years ago in a vain attempt to escape certain death.

Dick's thoughts were interrupted as the helicopter returned and the party jumped out to view the wreck. Someone suggested that they should pose for photographs, but this was done with little enthusiasm. The sense of sadness had spread. At that moment, more than at any time previously, Dick felt that he had an immense obligation to the dead aviators.

'We did not souvenir anything,' he said. 'Of course we could have easily thrust the *Kookaburra* into the back of our large truck and driven it to Sydney. It would have been years before anyone discovered. Until we relocated the *Kookaburra,* no one had been certain where it was, or even if it still existed.

'I decided to report the discovery to the Northern Territory authorities and discuss the future of the wreckage with them. We also sent a radio message via Wave Hill to the *Sydney Morning Herald,* in code, as arranged.

It said, 'The Night Parrot has been found'.

SEE PAGES 65 TO 68 FOR COLOUR PICTURES OF THE REMAINS OF THE KOOKABURRA AND KEY POINTS OF DICK SMITH'S EXPEDITION.

No rest for the Kookaburra

After searching the *Kookaburra* site, Dick Smith travelled to Wave Hill station and called the Northern Territory Chief Minister, Mr Everingham. He was not available, but his secretary advised Dick to telephone Dr Colin Jack-Hinton, the director of the Museum Board there. Dick did so, stating that although he intended complying with the law, he would provide exact details of the plane's location only if a number of conditions were met. He also asked if the Government could immediately provide a responsible person to protect the *Kookaburra* against 'souvenir hunters'.

Dick decided not to return directly to Sydney as planned, but to fly to Darwin and give the required information in person. He was about to leave Wave Hill station when a police party arrived to guard the plane. Dick gave them directions to the Hitchcock Highway, suggesting that they wait at the corner of the Cattle Creek fence, as anyone approaching the plane would proceed from there. The police insisted on knowing the exact location but, as he was worried about possible souvenir hunting, Dick replied with a position best described as approximate.

At Darwin, he met Dr Colin Jack-Hinton and advised him that he would provide exact location details on the following conditions:
(i) That the plane be retrieved by experts, preferably with archeological experience; and that it would not be out of Government hands at any time.
(ii) That it be moved in a pantechnicon large enough to carry the plane in one piece.
(iii) That the *Kookaburra* be displayed where as many people as possible could see it.

According to Dick Smith, Dr Jack-Hinton said he would be happy to comply, as he would want such conditions himself. He said he would supervise the recovery and, if necessary, borrow a suitable truck from the army. Dick offered to pay the cost of renting a pantechnicon if necessary. Further, Dr Jack-Hinton also said that the plane would be put on display in the foyer of an office building in Darwin.

Dick gave exact details of how to locate the plane. He wished to comply with the law, even though the fine for taking the *Kookaburra* out of the Territory ($200) was a trifle compared with the money which the search had already cost. However he felt that the law was unjust and that the plane should be returned to Sydney

— its real home — and displayed at Kingsford Smith airport. He repeated his point that the majority of Australians would never get the chance to see the *Kookaburra* in Darwin.

For some reason never explained, the Museum Board did not comply with his conditions. The plane was hurriedly removed on the back of a Toyota Land Cruiser by volunteers from Alice Springs, supervised by Dr

The *Kookaburra* as it was brought back to Alice Springs.

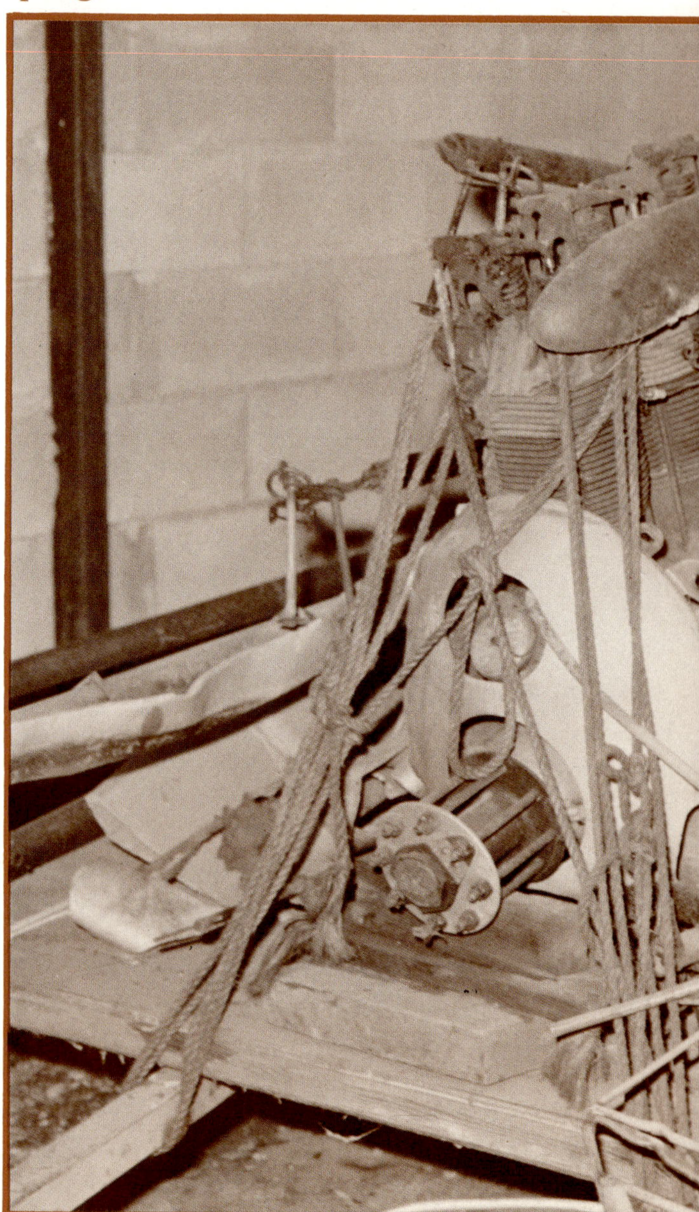

Jack-Hinton. He later admitted that some bolts had been removed with a hacksaw and some wing struts dismantled, but he said that the damage to the plane was negligible. In Alice Springs the plane was taken to the workshop of engineer and aviation enthusiast, Bill Kinsman, who had been in the team which retrieved the plane from the desert.

Dick Smith was horrified to learn that no plaque or other form of remembrance had been left to mark the spot where the *Kookaburra* had lain for 50 years.

He loudly voiced his disapproval and a controversy was soon raging about the 'right' place to display the historic aircraft. The president of the Aviation Historical Society of the Northern Territory, Mr Ian Silvester, announced that his society had no objection to Dick Smith restoring the *Kookaburra* provided that it was returned to Alice Springs. Charles Kingsford Smith — son of the famous aviator — joined the voiced opinion that the wreck should be returned to its 'home base', Sydney. The New South Wales Premier, Neville Wran and Keith Anderson's former fiancée, now Mrs Bon Tate, also requested that the plane be returned to Sydney.

Encouraged by this support, Dick Smith contacted Mr Paul Everingham, the Northern Territory Chief Minister with a proposal. He suggested that the *Kookaburra* be restored and rebuilt at Dick Smith's expense, but held in the custody of the New South Wales Museum of Applied Arts and Sciences and displayed at Kingsford Smith International Airport. In return, Dick Smith would undertake to make the plane available for display in the Northern Territory at regular intervals, once a suitable display venue was available.

The offer was rejected.

Part of the wreck of the *Kookaburra* is in the Aviation Museum at Alice Springs.

Postscript

As this book is being produced, parts of the *Kookaburra* are at different locations in Alice Springs, but no plans have been announced for its restoration. Its future is obscure but one fact is certain. Keith Anderson had a burning ambition to make headlines, but even he could not have foreseen that, 50 years later, his tiny Widgeon mono-plane would be the centre of such controversy.

On April 10th, 1979 — exactly 50 years after the *Kookaburra's* final flight — Dick and Pip Smith, their daughters, Garry Crapp and Tony Peter flew to Alice Springs in Dick's Beech Baron following the original route taken by Keith Anderson.

After attending a ceremony at which an official plaque was laid to commemorate the *Kookaburra's* final flight, Dick's party followed the exact route the *Kookaburra* had taken across the Tanami Desert and dropped a wreath over the site of the forced landing. Meanwhile John Haslett and his party had travelled from Darwin and made their way to the *Kookaburra* site. They held a ceremony in the turpentine scrub and laid an appropriate plaque.

During the proceedings, John Haslett stumbled across the *Kookaburra's* original tyre pump. Dick learned this and, shocked that the retrieving party had not scoured the area, he instituted his own search. He was rewarded with two buckets of 'bits and pieces' from the plane but the biggest surprise came like a gift from fate itself. In a small cleared area of ground, trapped in a clump of scrub, was a section of the original airframe fabric. It still carried the silver dope used when attached to the *Kookaburra* and somehow it had escaped the bush-fires.

Anderson and Hitchcock did not die in vain. An immediate effect on the original inquiry into the Coffee Royal Affair was that compulsory planning of flights was introduced. Certain parts of the country were designated 'remote areas' and pilots were permitted to fly over them only when carrying adequate radio equipment. Perhaps the most important outcome of the inquiry was the Federal Government's decision to have proper aeronautical maps made. Had it not been for the Coffee Royal Affair, many more aviators would have perished.

This alone ensured Keith Anderson and Bob Hitchcock a permanent place in Australia's aviation history.